The Man Who Would Be Queen

THE SCIENCE OF GENDER-BENDING AND TRANSSEXUALISM

The Man Who Would Be Queen

THE SCIENCE OF GENDER-BENDING
AND TRANSSEXUALISM

J. Michael Bailey

Joseph Henry Press
Washington, D.C.

Joseph Henry Press • **500 Fifth Street, N.W.** • **Washington, D.C. 20001**

The Joseph Henry Press, an imprint of the National Academies Press, was created with the goal of making books on science, technology, and health more widely available to professionals and the public. Joseph Henry was one of the early founders of the National Academy of Sciences and a leader in early American science.

Any opinions, findings, conclusions, or recommendations expressed in this volume are those of the author and do not necessarily reflect the views of the National Academy of Sciences or its affiliated institutions.

The names of some of the people mentioned in this book—and selected details about their lives—have been changed to protect the individuals' identities.

Library of Congress Cataloging-in-Publication Data

Bailey, J. Michael
    The man who would be queen : the science of gender-bending and transsexualism / J. Michael Bailey.
        p. cm.
Includes bibliographical references and index.
    ISBN 0-309-08418-0 (pbk. : alk. paper)
    1. Gay men—United States—Psychology—Case studies. 2. Transsexuals—United States—Psychology—Case studies. 3. Homosexuality, Male—Psychological aspects. 4. Transsexualism—Psychological aspects. 5. Gender identity—Psychological aspects. 6. Sexual orientation—Psychological aspects. 7. Nature and nurture. I. Title.
    HQ76.2.U5 B35 2003
    305.38'9664—dc21

                                2002154181

ISBN 0-309-08418-0

Printed in the United States of America.

For Drew/Kate

Contents

Preface   ix

PART I
THE BOY WHO WOULD BE PRINCESS   1

1 Princess Danny   3

2 Growing Pains   16

3 The Boy Who Would Not Be a Girl   39

PART II
THE MAN HE MIGHT BECOME   55

4 Gay Femininity   61

5 Gay Masculinity   85

6 Danny's Uncle   103

7 Is Homosexuality a Recent Invention?   124

PART III
WOMEN WHO ONCE WERE BOYS   139

8 Terese and Cher   145

9 Men Trapped in Men's Bodies   157

10 In Search of Womanhood and Men   177

Autogynephilic and Homosexual Transsexuals:
How To Tell Them Apart   192

11 Becoming a Woman   195

Epilogue   213

Further Reading   215

Index   221

Preface

The most expert cosmetics salesperson at the upscale department store in my neighborhood is a man. A female friend told me about him and, intrigued, I went to see him. He was young, tall, and African American, and his head was shaven. His fingernails were long and covered with clear nail polish. I watched him as he helped a woman choose the right makeup. After he was done with her, I introduced myself. He was slightly taken aback that I, a psychologist, wanted to meet him, but he also appeared slightly flattered. He told me his name was Edwin.

Knowing his occupation and observing him briefly and superficially were sufficient, together, for me to guess confidently about aspects of Edwin's life that he never mentioned. I know what he was like as a boy. I know what kind of person he is sexually attracted to. I know what kinds of activities interest him and what kinds do not. I am least sure what he will look like five years from now. Based upon his current appearance, there is a chance he will undergo a dramatic change.

Although I am virtually certain that my conclusions are correct, they fly in the face of mainstream academic opinion. If a current textbook discussed the basis of my intuitions—which many people share—it would do so in the context of stereotypes. It would neglect to explain that my intuitions are probably correct, and it wouldn't discuss why. My book aims to do better.

★★★★★★★★

Edwin is a feminine man, one of the most feminine men I have ever met. Any reasonable person who met him would agree with me, unless that person's only source of knowledge was a contemporary social science textbook. The textbook would say that concepts like "femininity" and "masculinity" are hopelessly muddled concepts that have more to do with the observer than the observed. Presumably its author would disapprove of using the word "feminine." It would be amusing to hear such a person trying to describe Edwin without it.

Scientifically, we have begun a renaissance period for taking femininity and masculinity seriously. This is partly because of men like Edwin, and partly because of boys like Edwin was. I do not ask Edwin about his childhood because I do not need to. I already know that Edwin played with dolls and loathed football, that his best friends were girls. I know that he was often teased by other boys, who called him "sissy." I am fairly certain that his parents did not encourage his feminine behavior, and if I had to bet, I'd say that his father was unhappy about it. The source of Edwin's femininity can be no obvious social influence. It might be a more subtle social influence, or it might be inborn. The fascinating question of what causes Edwin's femininity can be asked only if we admit that femininity exists.

★★★★★★★★

Although I didn't ask him, I know that Edwin likes to have sex with men. Not all gay men are like Edwin, but almost all men like Edwin are gay. During the past twenty-five years, social scientists have tried to discount or minimize the relation between male homosexuality and femininity. The standard lecture is that sexual orientation, gender identity, and gender role behavior are separate, independent psychological traits; a feminine man is as likely to be straight as gay. But the standard lecture is wrong. It was written with good, but mistaken, intentions: to save gay men from the stigma of femininity. The problem is that most gay men are feminine, or at least they are feminine in certain ways. A better solution is to disagree with those who stigmatize male femininity. It is a false and shallow diversity that allows only differences that cannot be observed.

To say that femininity and homosexuality are closely bound together in men may be politically incorrect, but it is factually correct, and it has been known for a long time. The idea that some males are "women's souls in men's bodies" was originally offered in 1868 to explain gay men, not transsexuals (by Karl Ulrichs, who was describing men like himself). Because the idea has been "off limits" among scientists for several decades, there is a host of fascinating phenomena well known to gay men and their friends that have barely been touched by scientists: the gay voice, the gay gesture, and prejudice against "femmes," to name a few. Scientifically demonstrating that these phenomena exist has been easy. The next step will be to try to understand why.

****

There is some chance that if I ever see Edwin again, his name and appearance will be changed to those of a woman. Even for a gay man, Edwin's appearance and manner are exceedingly feminine. He would stand out in a gay bar. (But he'd receive little romantic attention there.) He is near the boundary of male and female, and someday he may cross it. If he does, one primary motive will be lust.

The attempt to separate sexuality from gender has been especially

xii *The Man Who Would Be Queen*

misleading for transsexualism. Supposedly, male-to-female transsexuals are motivated solely by the deep-seated feeling that they have women's souls. Furthermore, the fact that some transsexuals are sexually attracted to men and others to women allegedly means that sex has nothing to do with it. However, in this case the exception proves the rule. Heterosexual men who want to be women are not naturally feminine; there is no sense in which they have women's souls. What they do have is fascinating, but even they have rarely discussed it openly.

One cannot understand transsexualism without studying transsexuals' sexuality. Transsexuals lead remarkable sex lives. Those who love men become women to attract them. Those who love women become the women they love. Although transsexuals are cultural hot commodities right now, writers have been either too shallow or too squeamish to give transsexual sexuality the attention it deserves. No longer.

★★★★★★★★

This book deals with feminine males and completely ignores masculine females. That was not my original attention. Butch women are fascinating too, and I have studied them. There are many analogies between very masculine women and very feminine men, but there are also important differences. Butch women are not simply the opposite of femme men. Rather than attempting to force them together, I decided to focus on males. Masculine females deserve their own book.

★★★★★★★★

Completing this book required substantial assistance from many other people. Several scientists and scholars spent a good deal of their time discussing ideas with me: Ray Blanchard, Khytam Dawood, Anne Lawrence, Simon LeVay, Rictor Norton, Maxine Petersen, Bill Reiner, and Ken Zucker. Anjelica Kieltyka introduced me to the Chicago transsexual community and taught me a great deal by being honest and open. My colleague, Joan Linsenmeier, read the entire manuscript and made sure that my thoughts were clear. My editor, Jeff Robbins, at Joseph Henry Press, made my writing better than I could. I am grateful to Daria Cooper for her support while finishing the book. Finally, I would never have thought of this book without Leslie Ryan and Cher Mondavi, both courageous women, in their own, different, ways.

The Man Who Would Be Queen

Part I

The Boy Who Would Be Princess

CHAPTER 1

Princess Danny

*I*t started with the shoes. After Danny Ryan became a proficient walker, not much more than a year old, he ventured into his mom's closet. He came out with a pair of strappy heels and struggled to put them on. Bemused, she helped him, and when he stood up in them, grasping her hand, he bounced with joy. This became something of a preoccupation for Danny. Often when he came into the bedroom, he went right for the closet. When it was closed, he pressed up against the door and whined. When she indulged him, he would pick out a pair of shoes, preferring the more feminine styles. One day, Danny came into the room with a sheet over his head and ran straight for the closet. He seemed more eager than usual to try on her shoes, and when he stood up in them and spread his arms, she was startled to realize, at last, the meaning of the sheet. It represented a dress. Danny was trying to dress like a girl.

Although Leslie Ryan felt intellectually satisfied with that simple explanation of Danny's behavior, she began to feel uneasy too. When she contemplated the reasons for her concern, she realized guiltily that she was falling prey to the same attitudes held by the bullies she loathed in junior high. It is surely common and harmless for children to explore the clothes and activities that society had assigned to the other sex. Why shouldn't they? Still, she found herself hesitating when Danny asked for help putting on her shoes. She encouraged alternative activities, such as reading or assembling puzzles or playing with the toys he was given. This tactic worked for a while, but invariably, he returned to the closet. She decided that she would neither encourage nor discourage his cross-dressing, as she had begun to call it.

However, when Danny's father, Patrick, first saw Danny in high heels clutching a purse, he did not share Leslie's tolerant attitude. He raised his voice: "Danny, get out of those shoes!" Danny liked neither his father's tone nor his message and, after a moment of stunned silence, began to cry. Leslie shot her husband a scathing glance and immediately picked Danny up to soothe him. Later, Danny's parents had a heated discussion. Danny's father said that it made him feel "creepy" to see Danny dressed as a female and thought that allowing him to do so set the wrong example. Patrick believed that parents are an important influence on whether a child becomes homosexual or heterosexual, and he wanted a heterosexual son. Leslie insisted that trying on female clothing at age 18 months could not make Danny a gay man, that children like to pretend to be lots of things, and that Patrick should just relax. She dared not tell him how often Danny was cross-dressing. Patrick's consulting job kept him on the road nearly five days a week, and when he was home, he was not the most attentive father. Leslie hoped that Danny's fascination with women's clothes would pass before his father had a chance to see it again. And for a while this seemed possible.

By age two, Danny had begun to follow his mother everywhere as she went about her daily routine, from cooking in the kitchen to dusting the living room, to talking on the phone to peeing in the bathroom. When she tried to get some time alone by turning on one of Danny's favorite videos (*The Little Mermaid* was his absolute favorite), Danny insisted that she watch with him. When other adults were around, Danny was particularly clingy. Once a friend brought over her rambunctious three-year-old son, and Danny was terrified of him. When the two boys were left alone together, Danny began calling "Mommy! Mommy!," ran to her, and buried his head between her legs. His mother did not remember Danny's older sister, Mary (now six and in school), being so afraid of being separated from her.

When Danny was about two and a half, he discovered his sister's room, with its dolls, dress-up clothes, pretend make-up kits, and especially, the tutu that she had long outgrown and that was only a bit too large on him. Mary rapidly lost patience with Danny's intrusions into her room and his fascination with her feminine things. She did not share her mother's reluctance to judge Danny's girlish behavior: "No Danny! Dresses are for girls. You are a boy." These altercations left Danny weeping in frustration and Mary furious, and so their parents framed the controversy in terms of territory and forbade Danny to enter Mary's room without permission. As a concession to Danny, his mother bought him his own Barbie doll and gingerly took his side when Mary criticized his feminine choices: "Danny can play with dolls if he wants to, as long as he stays away from yours. Everybody's different."

During the year after Danny's third birthday, his mother hired a regular babysitter for the afternoons in order to take an art history class. The sitter, Jennifer, was an attractive college student, a sorority girl who loved both children and fashion, and both Danny and Mary quickly idolized her. Leslie briefed her on Danny's unique behavior and reassured her that it was okay to indulge him. Soon Jennifer (at Danny's urging) was painting Danny's fingernails and letting him wear her bracelet. She introduced him to Barbie online, a website where they could dress up Barbie in an assortment of outfits. They also played "Princess Danielle," with Danny the princess and Jennifer the prince, wizard, king, or whatever male role the drama at hand required. Alternatively, they would produce sequels to *Aladdin* (with Danny playing the role of Jasmine), or *Beauty and the Beast* (with Danny playing Belle), or the latest video fascination with a beautiful female protagonist. Jennifer was amused to think that she had found a playmate so feminine that even she was relegated to the male role—and that this playmate was a boy.

It was about this time that Danny's parents had their second "Danny crisis." Patrick found Danny playing with his Barbie while wearing his sister's tutu, and furiously snatched the doll away. Then he picked up Danny, who was frightened, and carried him to the living room, where he accused Danny's mother "Look what your son is doing!" As she looked at their faces—Danny's ashen with fear and her husband's red with rage—Leslie felt her heart sink. She reached out for Danny, who practically leapt to her from his father and immediately began to cry loudly. She took him to his room and laid him on his bed, told him that she loved him and would be back in a little while, and returned to the living room to face her husband. In the ensuing discussion, she had to admit that Danny was cross-dressing regularly, but she thought that he was merely "going through a phase." She made her husband realize how devastated Danny was by the scene, and she saw his anger transform into guilty regret.

This was the last time they fought about Danny. After that day, the Ryans seemed to work out a silent compromise, in which Leslie tried to keep Danny's feminine side from her husband, and he let Danny alone. Danny helped, because he seemed to understand that his father was not as receptive as his mother to his feminine activities. Sometimes, despite their unspoken efforts, Patrick saw something not intended for his eyes—for example, Danny playing with Barbie. Although Patrick no longer stopped Danny or criticized him, these moments were usually awkward and tense. Danny would hesitate, as if he thought he might get in trouble, until Patrick left or looked away. Patrick would become cold and quiet, and Leslie would become especially attentive to him. But no one spoke up about Danny.

During Danny's fourth year, he "came out" to his block, going outside to play with the neighborhood kids, wearing or bringing whatever he wanted. Unlike his sister, the younger kids did not give him a hard time at first although some commented, "He's wearing girl's clothes." Danny gravitated toward the girls, who accepted him as a skilled participant in their activities, but he became visibly anxious when boys started playing rough around him, as they typically did. As the other boys began to shout, shoot each other with toy guns, and collide with each other, Danny shrank by the side of his guardian, usually Jennifer or his mother.

On his fourth birthday Danny had a party attended by several neighborhood kids and their mothers, his sister, Jennifer, and his mother. He wore his tutu, a bridal veil he had recently persuaded his mother to buy him, and black patent leather shoes that his sister had outgrown. Jennifer did his nails and fixed his hair (with a bow), and Danny was radiant. His gifts included a baseball bat and glove and a toy car (his mother and perhaps Danny too wondered why anyone would give these to him), some puzzles and books, a doll, a toy make-up kit, and, best of all, a charm bracelet from Jennifer. Leslie realized that the other mothers probably saw Danny as odd, but no one remarked about his outfit or the unusual gifts. Danny was ecstatic. He was on top of the world, happier than he would be for a long time.

****

Danny's fifth year was a turning point, the year of unhappy awareness. This was the year that other children, his mother, and Danny himself began to realize that his behavior was not only unusual, but also in some sense unacceptable. It was the year that Danny learned how cruelly our world can treat boys like him.

A new family with two older boys, ages 8 and 11, moved in down the street. The first time the boys met Danny, he was playing house with several girls. They studied him with increasing amazement before pronouncing him a "sissy" and a "fag," and they laughed at him derisively. Stunned, Danny ran home to his mother, who calmly explained to him the meaning of these words. Seeing the tears in his eyes as he struggled with the "sissy" accusation—after all, he *does* like girls' things—she was both furious at the boys and heartbroken for Danny. She realized that to that point, he had had an easy time with his femininity and that even if she could still intervene with these particular boys, there would be others with whom she couldn't. When she defined a "fag" to Danny as "a boy who loves other boys," Danny protested, "But I don't even like boys!"

Next time Danny went to play at the house where the boys had harassed him, his mother made sure she accompanied him. This prevented a repeat—the boys knew enough not to be mean to Danny in front of her—but it could not reverse serious damage to Danny's social situation. For one, she could not always be with him, and whenever Danny was on his own and saw his detractors, they made sure to tease him: "Fag!" or "Sissy!" or "Danny's gay!" or "Danny's a girl!" (Leslie thought it ironic that in other times and circumstances, the latter accusation would have made Danny happy.) Even more disturbing was the response of the other children, who began to question Danny's play preferences: "Danny, you shouldn't wear dresses. You're a boy." Occasionally, they rejected him outright: "You can't play with us." A couple of the older girls often protected Danny and scolded the others for picking on him, but the damage had been done. Even when Danny was allowed to play, there was now tension where before there had been none. Difference that had been ignored now mattered.

Leslie seethed with anger at the two boys who had spoiled Danny's world. She made an indignant phone call to their mother, who apologized, but nothing changed. She had violent fantasies of intimidating them into stopping. But she simultaneously realized that sooner or later Danny was destined to confront intolerance. Even if those boys had never clouded Danny's life, someone would have. Events soon proved her right.

Although Danny strongly preferred the company of girls, he had befriended one boy. Martin was not feminine like Danny, but he was on the quiet side, somewhat passive, and not rough. He didn't mind taking orders from Danny or playing the roles that Danny cast him in, which were, after all, invariably male roles. Danny's parents were pleased that he had found a male friend. Once Martin spent the night, and Danny spent several hours at Martin's one day the next week. Soon Danny asked his mother if he could have Martin over again. Leslie called Martin's parents to arrange something, and Martin's mother sounded strange as she said, "You need to talk to Martin's dad." Martin's father stammered a bit but otherwise sounded forceful as he explained: "We have a problem with the way that Danny plays. Last time he was here, he wanted to be the wife and he got Martin to play the husband. We don't think that's something our son should be a part of. So for now, I don't think that Martin and Danny should play together." She couldn't bear to tell Danny the truth and so told him that Martin was sick. When she told Danny's father later, she broke down sobbing to think of her son, four and a half years old, banned from his best friend's company.

It was becoming increasingly clear to Leslie that she would have to take a more active role in helping Danny negotiate his increasingly difficult world. She had never liked the idea of squashing Danny's feminine interests. Rather, she decided to help him become aware of the potential consequences of his choices. The outstanding issue, she decided, was cross-dressing in public. And so the next time Danny wanted to go outside wearing his tutu, his mother stopped him, "I don't think you should wear that, Danny."

"Why?"

"Because if you do, the other kids might be mean to you."

"But I want to."

"I don't think you should."

"Why?"

"I just told you. If you want to wear that, fine, but only in here. If you want to go outside, I want you to change into jeans."

By this time, Danny was crying and ran into his room slamming the door after him. She knew that she had hurt him, but what choice did she have? Could a four and a half-year-old boy reconcile the fact that there is nothing wrong with his strongest preferences with the conflicting fact that he must nevertheless hide them from most of the world? Could she reconcile these facts in her own mind?

Halloween approached, and she dreaded the unavoidable confrontation. When Leslie took her children to the store to get costumes, Mary chose Jasmine (from *Aladdin*). Initially, Danny tried to choose the same costume, but his mother said no. Danny thought her refusal meant that he should choose a different costume from his sister's. But when he selected a princess costume, his mother said, "I don't think that's best, honey," and suggested a cowboy costume. Disappointment flashed in his face, followed by shame. They eventually resolved that he would be a magician with top hat and cape and wand, but she had no illusions that this was Danny's first choice.

Danny asked for a bicycle for his fifth birthday, and they went to pick one out. Danny immediately chose a pink bike with streamers, and with Barbie painted on it. His mother said that this was probably not the best choice and tried to steer him toward a plain bike in blue or red or green. This time, however, Danny was in no mood to compromise. In the end, he chose not to get a bike at all rather than get one he did not want.

★★★★★★★★

Leslie became increasingly sad and worried about Danny and believed she was depressed. She decided to go to a psychiatrist who described himself as "psychodynamically oriented" and she told him all about Danny. The psychiatrist also wanted to know about her marriage, her own family, and her childhood. He seemed to focus on the period around the time when Danny was born, a period she had tried to forget. She had become depressed about her job. She found accounting unrewarding, but she had invested so much time and effort in taking courses and passing exams to be an accountant. Would she have to abandon her career goals to become a housewife? For the first year or so after Danny was born, she had been unenergetic and was not the attentive mother she should have been to either child. Gradually, she had accepted that being a homemaker to young children is a valuable job in itself, and that abandoning a career in accounting hardly made her a failure. Her energy returned, and she became a better mother to her children.

After a couple of months, the psychiatrist told her that he had reached an understanding of her case. He explained that Danny's feminine behavior was a direct consequence of her being unavailable to him during his first year—that because she was an absent mother, Danny had reconstructed a substitute woman in himself. Although he did not say so outright, it was clear that the psychiatrist believed that Danny's atypical behavior was all her fault. His primary recommendation was that she continue in psychotherapy with him, perhaps increasing to two visits a week. This feedback provoked a mixture of feelings in her. She had always felt guilty about her maternal behavior during this time and was now being confronted with the likelihood that, indeed, she had harmed her child. At the same time, something about the psychiatrist's formulation seemed a bit of a stretch to her. Can children really resort to such complicated solutions to their conflicts? At one year of age? And how is *her* psychotherapy going to help *Danny* cope with the intolerant reactions of other people?

She sought a school psychologist for a second opinion, thinking that a school psychologist had probably encountered boys like Danny and would have practical advice, especially because Danny was about to start kindergarten. She told their story to the school psychologist, who wrote a report. When the psychologist summarized the report to Leslie, she seemed more harshly judgmental than she had been during the interview. Giving Danny Barbies and letting him cross-dress were "inappropriate parenting behavior," she said. Danny's parents had been "neither willing nor able to set reasonable limits" on his feminine strivings, the report continued. The psychologist advised that if immediate steps were not taken, Danny faced social ostracism and would probably develop "a homosexual preference." Although this was certainly not the first time Leslie had considered his future sexual orientation, it was the first time that someone else had mentioned the issue so directly. She did not like the way the psychologist seemed to assume that homosexuality would be a bad outcome. In her own mind, the issue was more complicated—she wanted Danny to be happy, and if he could be both happy and gay, she would love and accept him all the same. And compared with Danny's current predicament, homosexuality seemed a minor consideration. By the time the school psychologist finished presenting her report, Leslie was in tears. Noticing this, the psychologist said: "I understand that this is difficult for you to hear, but we both want what is best for Danny."

Around this time, Leslie learned something about her family that she felt must be relevant. Her 40-year-old brother, Mark, called to say that he was divorcing his wife because he was gay. Mark said that he had recognized homosexual feelings in himself from childhood and had had sex with men beginning in adolescence and even through his marriage. But he had felt "Catholic guilt" and tried, at least intermittently, to suppress his gay feelings. After falling in love with a man, Mark realized that he could never be happy unless he followed his heart, and this required self-acceptance. Leslie was stunned by his revelation but managed to reassure her brother that it would not hurt their relationship. Later, discussing this with her own mother, she made a connection. She asked her mother whether Mark had been a feminine boy. Her mother, who knew about Danny, revealed that, indeed, he had been. When Mark was very young, he liked dolls and even cross-dressed a couple of times. Their father had disliked these behaviors and wouldn't allow them. He had worried that Mark was becoming a sissy and made him play sports, which Mark detested. She hadn't mentioned this before because Mark seemed ashamed to be reminded of his feminine past, and so she didn't want to bring it up. She had always assumed that because Mark "outgrew" this behavior, Danny would also. Leslie could not believe that the similarity between Danny and Mark was merely a coincidence, but if not, what did it mean? Was there something about their family that produced feminine boys? A gene perhaps? Would Danny become a gay man, like Mark?

****

Kindergarten started off well enough. Leslie met with Danny's teacher a week before school began. She said that Danny was "special," and then explained how. The teacher insisted that she would not permit other children to give Danny a hard time, and her attitude was confident and reassuring.

Leslie also talked to Danny, in order to prepare him. As delicately as she could, she suggested that Danny not talk to the other children about "girl stuff" for the time being, that Danny shouldn't bring dolls or girls' toys to school. She took Danny to Nordstrom to pick out clothes for the first day of school. With her guidance, they selected a red Ralph Lauren polo shirt (with the polo logo), navy khakis (pleated and cuffed), and black tasseled loafers. Danny looked proud when the salesman said, "You're going to be the best dressed boy in your school." They had a bit of a conflict about Danny's lunchbox. He went straight for a lavender Aladdin and Jasmine number, but she sadly refused. She saw a blue one, featuring Aladdin and the genie, and Danny objected, disappointed that there was no Jasmine. They settled, eventually, on a red Aladdin version, sans Jasmine.

Leslie had been dreading the moment when she dropped Danny off at class the first time. He had always been unusually attached to her, and this separation would be for several hours in a new, potentially scary, environment. They had discussed various scenarios that concerned him: what if he got sick at school, what if she forgot to pick him up, what if he got lost in the building. But at the moment of truth in early September, Danny gave her a quick hug, said "Bye," and marched in. Leslie watched for a moment, then turned on her heel and rushed out of the room so that Danny wouldn't see her cry.

The first day was a short one, for orientation purposes, and three hours later she anxiously re-entered the room and Danny ran to her smiling. She felt immensely relieved. Maybe this could work. Maybe it was even good for Danny. And for a while, it seemed so.

Then one afternoon about a month later, Leslie was called to pick Danny up from school early. He had had a problem during recess, and he had been crying nonstop ever since. When she picked him up, he fell into her arms, and he couldn't stop sobbing long enough to explain what had happened. She took him home, and all he wanted to do was sit in her lap quietly and watch television, periodically wailing and crying, while she soothed him, quietly insisting that everything was okay. Eventually, he calmed down enough to tell her what had happened.

He had been playing with some girl friends on the playground. Suddenly, a group of boys swooped in, shoved him to the ground, and for good measure, a large one jumped on him, knocking the wind out of him. Leslie could imagine just how he felt, because she believed she felt the same way: betrayed for no apparent reason, with no warning. She wondered if this was some kind of random careless act or if they had specifically targeted Danny. She also wondered how Danny's behavior had registered with the other kids. His teacher had seemed sympathetic but surprised at how Danny had reacted. The next morning before dropping him off, she assured Danny that he would be safe. That afternoon, she learned that things had only gotten worse. At recess three boys followed him around calling him "girlfriend," "fairy," and "faggot," until he latched onto the teacher and she scolded them.

Leslie arranged to talk to the teacher, who said she was angry with the offending boys and promised to protect Danny. But she added: "I'm concerned that Danny is doing some things that make other kids dislike him. He's bossy and demanding. He tells the girls he plays with what they have to do and say. He tattles. And when other kids tease him, instead of ignoring them, he talks back to them in ways that egg them on. The other day, someone called him a girl, and he said: 'I'd rather be a girl than a stupid ugly boy.' On the one hand, I admired his chutzpah, but on the other hand, I knew that this would only make things harder for him." Leslie's hopes for Danny's easy adjustment to kindergarten were destroyed.

Danny no longer wanted to go to school, but his mother managed to get him there anyway. She considered any uneventful day a good day. Whenever anything happened, it was usually bad. He had become an outcast at school, and he also seemed to enjoy his life outside of school less. She worried that he was depressed. When she raised the possibility of taking Danny to a therapist ("to talk to about things that bother you"), he initially resisted. He assumed that the therapist would want to talk only about his femininity, and he was ashamed and defensive. But his mother reassured him that he could talk about whatever he wanted, and that she didn't want to change him. He eventually agreed to see a child psychologist, who in turn gave Leslie the name of a therapist she could see for the depression that she felt returning.

Current events gave Leslie one more concern. A teacher in one of the wealthy suburbs made all the newspapers because he ended one school year as a man and began the next as a woman. Danny found out—one of the kids at school told him that he should follow suit—and was *very* interested. ("What happened to his penis? Can she have babies? Is she pretty?") Transsexualism had always been in the back of Leslie's mind, albeit distantly, but Danny's reaction made her more anxious. Even if she could handle Danny becoming gay, the possibility that he would get surgery to become a woman was not something she could tolerate.

CHAPTER 2

Growing Pains

*I*n spring of 1996 Leslie Ryan came to my Northwestern University office to seek yet another opinion. Jennifer, Danny's sitter, was a student in my human sexuality class and was working in my laboratory on studies of sexual orientation. I had lectured in class and spoken in lab meetings about feminine boys, and Jennifer thought that I might be able to give Leslie more definitive answers than those she had obtained thus far. Danny's mother had three general questions: Most importantly, What is the best way to raise feminine boys to be happy boys? For the sake of curiosity, Where do boys like Danny come from? For the sake of both curiosity and helping Danny, What becomes of feminine boys? I could easily answer only one of her questions. I have a good idea what Danny will be like when he grows up.

★★★★★★★★★

Leslie insisted to me that she would love Danny no less if he grew up to be gay. At the same time, she was curious whether he would, and she also realistically thought that his life would be more difficult if he were gay. And she knew that Danny's father desperately wanted Danny to grow into a heterosexual man.

Many people believe that feminine boys become gay men. When Danny was only three, the Ryans had discussed the possibility that he might become gay. Of course, the children who had tormented Danny—calling him "fag"—were already convinced that Danny was gay. That was undoubtedly why they wanted to torment him. Most people Leslie has confided in have also broached the issue with her. They seemed to be divided between two general opinions. Some people recognized that the belief that feminine boys become gay is a stereotype and so rejected it the same way they rejected most stereotypes, which, they felt, are the product of unenlightened thinking. Others wondered if there might well be something to the idea—just as many people speculate (most often in private) about the truth of other stereotypes. Social scientists have studied what becomes of boys like Danny, and it is the one question about the boys that they have effectively answered, one area in which even responsible social scientists can give an answer that is more than a highfalutin' way of saying "I don't know."

Several scientists have followed nearly 100 feminine boys from childhood into early adulthood. Because of their work, we can make educated predictions regarding Danny's adult sexuality. Most likely, Danny will become a gay man. It is also possible, although less likely, that he will grow up to be heterosexual. The final possible outcome is that Danny will decide to become a woman, and in this case, he will also be attracted to men.

The largest, most famous, and best study on this issue was conducted by Richard Green, then a psychiatrist at UCLA. Green began with 66 feminine boys, mostly referred by therapists. He also recruited a control group of 56 typically masculine boys. The boys' average age was about seven years old when Green first saw them, although some were as young as four and others were as old as twelve.

The feminine boys exhibited a variety of feminine behaviors:

• Cross-dressing: Nearly 70 percent did this frequently, compared to none of the boys in the control group.
• Playing with dolls: More than 50 percent did this frequently, compared to less than 5 percent of the control group.
• Taking female roles in games such as playing house: Nearly 60 percent took the female role, compared to none of the control group.
• Relating better to girls rather than boys as peers: About 80 percent did so, compared to less than 5 percent of the control group.
• Wishing to be girls: More than 80 percent stated such a wish occasionally, compared to less than 10 percent of the control group.
• Having below-average interest in rough-and-tumble play and sports participation: Nearly 80 percent had below-average interest, compared to 20 percent of the control group.

These were clearly two very different groups of boys, and the feminine group was on the extreme side. Danny showed 5 of the 6 behaviors; he has never expressed outright the wish to be a girl.

The boys' parents reported that their sons' feminine behaviors emerged quite early. For instance, more than half of them said that cross-dressing had begun before age 3 and virtually all cross-dressing began by age 6. Parents varied considerably in their initial reactions to the feminine behavior. Some parents were horrified and intolerant. Others seemed to have found the behavior cute, at least at first. They showed Green photographs of their sons wearing high-heeled shoes and dresses, and they admitted that they had bought their sons dolls. Mothers who remembered reacting more positively had sons who were slightly more feminine at the time Green first saw them. However, this effect was small, and one wonders how much mothers' memories might be biased by their sons' present behavior. Leslie's initial reaction was neither positive nor negative. Although emotionally she was more concerned than delighted, her overt response to Danny was to tolerate his femininity, if not to encourage it.

Green tried to stay in touch with the boys as they became teenagers and adults. At the final follow-up he collected data from about two-thirds of the boys—it is practically impossible to maintain contact with subjects in long-term studies such as Green's. On average, the boys were 19 years old during their final interviews, the youngest being 14 and the oldest 24.

The results of Green's study are among the clearest and the most striking in all of developmental psychology. About 75 percent of the young men who had been feminine boys said that they were attracted to men, compared with only one young man who had been a typical, masculine boy. The odds against these results being due to chance are astronomical.

The other 25 percent of the young men who had begun as feminine boys denied attraction to men. Green does not seem very skeptical about these denials, but I am. For one, the 25 percent who claimed to be heterosexual were three years younger, on average, than the 75 percent who admitted attraction to men. Coming out as gay to others, or even to oneself, sometimes takes time, and it is likely that at least some of the 25 percent who claimed to be heterosexual would eventually become gay men. Green himself wrote of some subjects who denied homosexuality at earlier ages and then admitted later that they had not been completely honest. It is conceivable that every one of the feminine boys grew up to be attracted to men. I am not arguing strongly that this is true—we simply do not know.

At his final interview, Todd, one of the young men from the feminine group, said he wanted to become a woman. Nothing clearly distinguished Todd's childhood from that of the other feminine boys. His parents reacted negatively to his femininity. His father, in particular, was angry about it, sometimes telling Todd to stop and sometimes ignoring his cross-dressing and playing with dolls. At puberty, as Todd began to mature physically, he realized that he wanted to do so in the female direction. He was somewhat small for his age. At age 17 he said that he wished he had breasts and a vagina and, although he knew it was impossible, wished he could give birth. He was attracted only to men. At his final interview at age 18, he said that his mother had given him a book about Christine Jorgensen, the first person ever to have a sex change operation, and he had become obsessed with it.

We don't know whether Todd ever became a woman, but let's assume he did. It might seem that if only one of the feminine boys grew up to be transsexual, then being a feminine boy is not very strongly related to adult transsexualism. But transsexualism is a very rare outcome; in Western countries, only about 1 in 12,000 males undergoes a sex change. Even if Todd was the only one, the rate of transsexualism among the feminine boys was about 400 times higher than we would expect in the general population. And conceivably some of the feminine boys Green lost touch with became transsexual—feminine boys who become transsexuals are often estranged from their families and so are more difficult to contact. Some other scientists believe that Green's transsexualism rate was on the low side, although no one believes that transsexualism is nearly as common an outcome as homosexuality is.

✶✶✶✶✶✶✶✶

When I told Leslie about the prospective data on boys like Danny, she said that she didn't care that he will probably become gay, that she only wanted him to be happy. I believed her, but in any case, her attitude is sensible. There is no reason to believe that we could alter Danny's future sexual orientation even if we tried. Several boys in Green's sample were treated for their feminine behavior, sometimes by therapists who believed that homosexuality would be a bad outcome. But the rate of homosexuality among the treated boys was no different than among the others.

Still, she worried how she should act toward Danny, what would help him have the happiest possible life. Should she accept his feminine tendencies completely and indulge his atypical desires? Should she have bought him the Barbie bike? Or should she do the opposite, firmly and consistently discourage the behavior that has led him to ostracism? Should she even discourage his private sex-atypical behavior—throw out his girls' clothes, for example?

Increasingly, Leslie felt torn. When she tolerated Danny's girl-like behavior, she did so uncomfortably, wondering whether she was being overly tolerant. After all, children don't get to do everything they want to do. They don't get to eat candy, stay up late, or stay home from school whenever they want. Was she failing Danny by not setting firm limits on behavior that was ultimately self-destructive? But when she did set limits, she felt more than just discomfort. When she saw the disappointment, anger, and shame in Danny's eyes, invariably followed by tears, she felt heartbroken. At those moments she wanted to tell him that she loved him just as he was, that he should do whatever made him happy, that she would always protect him from the reactions of others. But she knew this was impossible.

If she knew that in the long run Danny's happiness would be maximized by the short-term misery of squashing his femininity, she could do it. Or if the opposite were true—that Danny would be happiest if allowed to flourish in his own way, and that preventing this would only damage him—her inner conflict would cease; if she only knew what to do.

Unfortunately for Leslie, psychologists don't always know what is best, and we probably will not know for the foreseeable future. However, it is conceptually simple to design a scientific study to answer the question. First, identify a group of boys like Danny. Next, randomly assign them to be treated differently, with half the boys being indulged and the other half discouraged in their femininity. Follow them into adolescence and on to adulthood, and see if they differ in their outcome. However, besides taking years to complete, such a study would require that parents be indifferent to having their feminine boys assigned to either of two radically different treatments, with the possibility of harm. (Of course, parents' actions may already be harming the boys, but at least the parents themselves are choosing how to treat them.) It would also require serious research funding, to pay therapists and researchers, but the issue has become the kind of ideological battleground that funding agencies do not like to touch.

So I do not know what to tell Danny's mother about the best way to treat Danny. I can only tell her what several experts, who have studied and treated boys like Danny, recommend, and why. Unfortunately, the experts disagree among themselves, some of them passionately so. Indeed, the controversy concerning what to do about children like Danny has become one of psychiatry's hottest potatoes.

****

According to the current *Diagnostic and Statistical Manual of the American Psychiatric Association* (*DSM-IV-TR*)—which represents a kind of official list of mental disorders—Danny has a mental illness: childhood gender identity disorder (or GID for short). "Gender identity" refers to the subjective internal feeling that one is male or female. Most of us rarely, if ever, think about our gender identities. But if we imagined that others were treating us as the opposite sex—insisted that we were the opposite sex—most could get an idea of the mental anguish a child with GID may feel.

To be diagnosed with GID, a boy must meet four major criteria. (These are similar criteria to those for GID girls, although obviously, girls and boys with GID behave nearly oppositely.) First, he must behave in very feminine ways. Second, he must show signs of being unhappy as a boy. Third, his life must be substantially and negatively affected by his symptoms. Fourth, his atypical behavior cannot be due to a known medical syndrome that interferes with sexual differentiation, or the process of becoming male or female. (One example of this would be Kleinfelter's syndrome, a condition in which boys are born with an extra X chromosome.)

The controversy focuses on the first two criteria, and particularly on the second. So let's look at them more closely. In order to meet the first, behavioral, criterion a boy must show at least four of the following:

- A repeatedly stated desire to be, or insistence that he is, a girl
- A preference for cross-dressing or simulating female attire
- Strong and persistent preferences for female roles in make-believe play or persistent fantasies of being female
- An intense desire to participate in stereotypically feminine games and pastimes
- A strong preference for female playmates

The second major criterion concerns feelings, and in particular gender dysphoria, or discomfort with one's biological sex. Children are not very articulate about their feelings, and so we often infer their feelings indirectly. The *DSM* gives a range of behaviors that can provide evidence of gender dysphoria. In boys, the most extreme forms of gender dysphoria include the wish not to have a penis. But a boy can also pass the gender dysphoria hurdle if he shows "aversion toward rough-and-tumble play and rejection of stereotypically male toys, games, and activities."

Regarding the behavioral criteria, Danny has at least 4 of the 5 behaviors. (A few times when he was younger, he playfully insisted that he was a girl. This doesn't qualify as a "repeatedly stated insistence.") Regarding gender dysphoria, he has never complained about his penis, but he certainly dislikes rough-and-tumble play and rejects stereotypically male activities. Danny is not even a close call, diagnostically speaking.

The current controversy in the mental health professions regarding what to do with boys like Danny is strongly related to attitudes toward the GID diagnosis. Some experts think that it is obvious that boys like Danny have mental problems that need to be treated. In contrast, an emerging group of mostly (but not entirely) gay thinkers believe that the childhood GID diagnosis should not exist. They believe that the diagnosis does far more harm than good. The two groups of experts would give very different recommendations to Danny's mother.

Leslie knows about GID, and she unhesitatingly rejects the idea that Danny is mentally ill. But that does not resolve her dilemma, nor does it ease her mind. Danny is not mentally ill because he is feminine, but he is having problems and is too often unhappy, and she does not know how to help him.

✶✶✶✶✶✶✶✶✶

One approach that some clinicians have taken to boys like Danny is socially conservative. Its most visible advocate is George Rekers, who is a member of the ultra-conservative "Leadership U," a virtual university on the Web. Rekers is an academic psychologist who held positions at Harvard and the University of California, Los Angeles, before assuming his present position at the University of South Carolina. He has published numerous academic articles and several books, and at one point he was funded by the National Institutes of Health to research the treatment of children with GID. Yet there are disturbing aspects of Rekers' work that are peculiarly unscientific, such as his writings invoking religious arguments for the superiority of heterosexuality. His assertion that homosexuality is "an unfortunate perversion" was, even in 1982, certainly out of fashion in academia, which tends to be socially liberal. Rekers represents the right wing of gender theorists and therapists.

Rekers' position seems to be essentially the same as that of the National Association for Research and Therapy of Homosexuality (NARTH): homosexuality is inferior sexuality; homosexual people can sometimes be successfully changed into heterosexual people; homosexuality is the result of faulty learning and abnormal family dynamics, so the earlier the intervention the better; feminine boys are sick and at risk for homosexuality; feminine behavior can be eliminated the way that many other undesirable behaviors can, by consistent application of reward and punishment.

Rekers has published vivid case histories of some of his patients, and perhaps the most interesting was known by the pseudonym "Kraig." Kraig was especially important because he was a member of Green's long-term study (Green named him "Kyle"), and so we know something about how Kraig turned out.

Kraig entered therapy when he was about five years old. Both his parents were quite worried about, and his father was in fact intolerant of, Kraig's feminine behavior. At least once prior to therapy his father spanked Kraig for putting female clothes on his stuffed animals. Kraig's therapy involved the application of behavior-modification principles that are familiar to many psychologists, teachers, and parents. For example, Kraig's mother was trained to ignore him whenever he displayed feminine behavior. This was initially quite traumatic for both of them. Kraig screamed so loudly in the laboratory during one session that he had to be removed by a laboratory assistant. Kraig was also put on a "token economy," in which he was given different colored tokens for masculine and feminine behavior. The blue tokens he earned for masculine behavior could be exchanged for treats such as candy bars. The red tokens he earned when he was "bad"—feminine—had bad consequences ranging from loss of blue tokens to loss of television time to the most effective punishment: being spanked by his father. Although training occurred in the laboratory, these techniques were applied in all areas of Kraig's life, for example, including his choice of male versus female playmates.

According to Rekers, after 60 sessions Kraig engaged exclusively in male-typical behavior. Rekers' treatment team noticed, however, that in the laboratory Kraig seemed to be acting. He would approach the table of toys and say something like "Oh look at those girls' toys. Yuck. I don't want to play with those. Where are the good boys' toys?" Still, Rekers convinced himself that Kraig was a clear success. Indeed, two years after treatment ended, his mother was concerned that Kraig had become too rough and destructive. Rekers advised her that this problem also was treatable and was preferable to the excessive femininity that Kraig initially displayed.

Green saw Kraig periodically between the ages of 5 and 18. When Kraig was 17, his mother was interviewed and said she was thankful that he had had the therapy; that without it he would have doubtless become homosexual or worse. Unfortunately, however, the therapy had not rescued Kraig's relationship with his father, which had only gotten worse. (It seems that Kraig never learned to enjoy hunting with his father and preferred art and theater to sports.) At age 17 Kraig was telling a story similar to his mother's, indicating his disgust with homosexuality and men who behaved in a feminine way. A year later, however, Kraig admitted that he not only had homosexual feelings but that he had acted on them—with a complete stranger in a restroom in a convention center. He felt in his mind that the experience was "unreal," and shortly afterwards took an overdose of aspirin. (He survived.) He believed that his parents would be disappointed and upset if they found out that he was not heterosexual. In general, Kraig appeared to be ashamed and deeply conflicted about his homosexuality. But he no longer enjoyed dressing or acting like a girl.

****

Opposite Rekers on the gender political spectrum is a group of increasingly vocal clinicians, writers, and theorists, who believe that boys like Danny are healthy victims of a sick society. They include psychiatrists Richard Isay, Ken Corbett, and Justin Richardson, psychologist Clinton Anderson, scientist Simon LeVay, and journalist Phyllis Burke, author of *Gender Shock*. (All these individuals are homosexual, but this movement also includes some heterosexual supporters.) They argue that there is nothing inherently wrong with children who behave like the opposite sex. Most of these writers accept that there is a strong correlation, in boys at least, between early sex-atypicality and later homosexuality. (Burke is an exception.) Because homosexuality is normal and healthy, feminine pre-homosexual boys should not be labeled sick any more than gay men should be. (This argument is closely analogous to the one that Rekers uses to generate the conclusion that homosexuality is a form of mental illness.) The problem that feminine boys face is that of an ignorant, intolerant society, a society that allows people to be cruel to them for no good reason. Treating a man or boy for "femininity" does more harm than good.

Isay, for example, says that virtually all the gay men he has seen in his clinical practice had some feminine traits in boyhood, and a few of them were sent by their parents for psychological treatment. According to Isay, his gay patients who had been treated during childhood for being prone to cry easily were now uncomfortable with emotional expression. Treated for excessive femininity, they now tried to distance themselves from all things feminine, despite the fact that femininity is part of "their nature." The result can only be unhealthy inner conflict.

LeVay and Burke both point to Rekers' patient, Kraig, as a kind of poster child of the harm that the GID notion produces. To them, the primary results of Kraig's treatment were damage to his self-esteem and the crippling of his ability to express his romantic and sexual feelings toward men.

A recurring theme among the critics of the childhood gender identity diagnosis is that it includes children who simply do not conform to stereotypes of the other sex, whether or not the children have deeper gender identity problems. In other words, a boy who acts like a girl but is happy being a boy could still earn a diagnosis of GID. Although gender dysphoria is ostensibly a core component of the syndrome, in order to meet the criterion it is sufficient that a boy avoid typical boys' activities. This would make sense if boys who strongly preferred acting like girls to acting like boys invariably did so because they wanted to *be* girls and disliked being boys. They think that deep inside, Danny Ryan wants to become a girl, whether or not he says so. The critics of the childhood gender identity diagnosis believe that, often, feminine boys just like to be feminine boys, and no more. Similarly, most of these critics downplay any association between symptoms of childhood GID and later transsexualism. They do not believe that the femininity of boys like Danny implies a fundamental gender dysphoria that typifies transsexual adults. They think that Danny Ryan just likes to act like girls do, but that he would be content being a feminine boy.

The anti-GID folks have a logically consistent treatment recommendation: no diagnosis, no treatment. They do not believe that Danny needs psychotherapy to help him become more masculine or satisfied with being a boy. Rather, they believe that most boys with GID—even boys who declare that they are girls—will grow out of it on their own. And they are uniformly horrified by the behavioral techniques applied by Rekers. To be sure, they do not think that boys with GID have easy lives, and they do not believe the boys should be ignored. Rather, they want to change society so that feminine boys are treated less badly. I initially was quite skeptical about this position, because it seemed to smack of ideological grandstanding at the expense of feminine boys. Who can really hope to change society? I once challenged LeVay on this, and he told me about a teacher friend of his who had a GID boy in class, and who helped the class come to terms with the boy's "odd behavior and appearance." And he reminded me of how dramatically some other societal beliefs had recently changed. So I became less skeptical, if not yet convinced.

★★★★★★★★

Ken Zucker, head of the Child and Adolescent Gender Identity Clinic in Toronto, has criticized the right-wing position on GID, and he has recently clashed with the left-wing. Therefore, and perhaps because I find his position especially balanced, I consider him to be the moderate. With his thick gray beard and his contemplative manner, Zucker appears rabbinical. He has certainly acquired a Talmudic knowledge of the literature concerning childhood GID, about which he is the world's leading expert. His book *Gender Identity Disorders in Children and Adolescents* is surely the most comprehensive text ever written on this topic. To say that Zucker is knowledgeable is an understatement. To say that he is obsessive about certain subjects, including GID and sexual orientation, is only a slight overstatement. I have seen Zucker in academic action for a number of years. He has reviewed my articles, for example, and now he is editor of the prestigious journal, *Archives of Sexual Behavior,* and decides whether articles should be published there. Invariably, he has pointed out several mistakes in my papers, from omissions of prior research I had been unfamiliar with, to punctuation mistakes in my reference list (!). For this reason, I tend to give Zucker the benefit of the doubt in certain respects. For example, I do not think he is prone to make mistakes due to being uninformed or rushing to decisions. This is not to say that he is always right.

So what is Zucker's position? First, he believes that the diagnosis of childhood GID is useful and valid, and the diagnosis is not merely a value judgment that boys who like girls' activities (or girls who like boys' activities) are sick or wrong. This is due to his conviction that children with GID suffer, and that the suffering is not only attributable to bullying by closed-minded peers and adults. Second, Zucker thinks that kids with GID often need to be treated with psychotherapy, and that their families do as well. These beliefs obviously distinguish Zucker's opinion from that of the left—"leave masculine girls and feminine boys alone"—crowd, but Zucker also disagrees with the right's emphasis on preventing homosexuality. Zucker does not consider this an important clinical goal, because he thinks that homosexual people can be as happy as heterosexual people, and regardless, he doubts that therapy to prevent homosexuality works.

However, when I spoke to Zucker about the current debate about childhood GID, I came away with the impression that these days, he feels besieged primarily on the left. He has had several recent exchanges in academic journals on the issue of GID, all with critics who believe that the GID diagnosis is essentially gender repression; his tone in some of these exchanges has seemed irritable. He has argued, among other things, that the notion that a boy might be diagnosed simply for liking dolls is completely wrong. All the children referred to his clinic for GID have had significant cross-gender behavior.

More importantly, he has scoffed at the idea that children with GID are unhappy only because they are socially ostracized. He remembers cases in which children were unhappy primarily because they couldn't become the other sex. For example, he recalls parents of a boy with GID telling him: "Every night before going to bed, he prays to God to turn him into a girl." Another mother of a six-year-old boy with GID told Zucker that the boy cried himself to sleep every night, softly singing, "My dreams will never come true." These boys are unhappy because they aren't girls, regardless of whether others call them "sissy." Zucker thinks that an important goal of treatment is to help the children accept their birth sex and to avoid becoming transsexual. His experience has convinced him that if a boy with GID becomes an adolescent with GID, the chances that he will become an adult with GID and seek a sex change are much higher. And he thinks that the kind of therapy he practices helps reduce this risk.

Zucker emphasizes a three-pronged treatment approach for boys with GID. First, he thinks that family dynamics play a large role in childhood GID—not necessarily in the origins of cross-gendered behavior, but in their persistence. It is the disordered and chaotic family, according to Zucker, that can't get its act together to present a consistent and sensible reaction to the child, which would be something like the following: "We love you, but you are a boy, not a girl. Wishing to be a girl will only make you unhappy in the long run, and pretending to be a girl will only make your life around others harder." So the first prong of Zucker's approach is family therapy. Whatever conflicts or issues that parents have that prevent them from uniting to help their child must be addressed.

The second prong is therapy for the boy, to help him adjust to the idea that he cannot become a girl, and to help teach him how to minimize social ostracism. Zucker does not teach boys how to walk in a manly fashion, but he does give them feedback about the likely consequences of taking a doll to school.

The third prong is key. Zucker says simply: "The Barbies have to go." He has nothing against Barbie dolls, of course. He means something more general. Feminine toys and accoutrements—including Barbie dolls, girls' shoes, dresses, purses, and princess gowns—are no longer to be tolerated at home, much less bought for the child. Zucker believes that toleration and encouragement of feminine play and dress prevents the child from accepting his maleness. Common sense says that a boy who wants to play with dolls so much that he is willing to risk his father's wrath and his peers' scorn is unlikely to change his behavior due to inconsistent feedback, sometimes forbidding, sometimes tolerating, and sometimes even encouraging it. Inconsistent parenting like this is ineffective in stamping out any kind of unwanted behavior.

Compared with the therapy of the right-wingers, Zucker's therapy is more psychologically focused and less punitive. Although Zucker encourages parents of GID boys to set limits on their sons' feminine activities, he also encourages parents to discuss their gender concerns openly with their sons. Still, there is no denying that both moderate Zucker and right-winger Rekers think that parents should not just sit back and let their sons express their feminine sides. This view draws the wrath of the left-wingers, who insist that there is nothing wrong with boys who like girls' things. The central difference between Zucker and his critics on the left is that Zucker believes that most boys who play with girls' things often enough to earn a diagnosis of GID would become girls if they could. Failure to intervene increases the chances of transsexualism in adulthood, which Zucker considers a bad outcome. For one, sex change surgery is major and permanent, and can have serious side effects. Why put boys at risk for this when they can become gay men happy to be men?

I have not heard anyone argue that transsexualism is an acceptable outcome for feminine boys. This possibility is worth thinking about, though, and in a moment I will. For now, let's assume that we don't want boys to become girls and consider whether Zucker's methods are necessary. One leftist, the scientist Simon LeVay, has argued that most boys with GID grow up as normal gay men without therapy, and so the discouragement of femininity that Zucker recommends is unnecessary and even cruel.

One bone of contention is the rate of untreated boys with GID who would become transsexual. Because Richard Green's prospective study is so famous, it is common for people to cite his transsexual outcome rate of 2 percent (one boy out of fifty). However, a more comprehensive review found a rate of 6 percent, and the authors (Zucker was one) believe that this may have been an underestimate. Transsexual adults are more likely than gay men to be estranged from their families and unable to be found for a follow-up study. So maybe transsexualism is a more common outcome than some people believe.

Recent experience in the Netherlands, where attitudes towards transsexualism are quite liberal, also suggests this. Peggy Cohen-Kettenis, a psychologist at the gender clinic for children and adolescents at the University Medical Center in Utrecht, recently reported that 23 percent of the clinic's adolescent patients who had been evaluated for gender identity problems as children went on to request sex reassignment when they became eligible to do so at age 12.

Still, most boys who want to be girls become men who don't want to be women. In the Zucker-LeVay exchange, LeVay didn't say how he thinks this happens. But he did imply that it is unnecessary to try to make boys with GID more like other boys. Somehow, perhaps through psychological maturity alone, they will lose their desire to be girls and their unhappiness to be boys. The problem with this analysis is that it ignores what happens in the lives of these boys, even those who get no therapy. In contemporary America (and in every other culture I know), very feminine boys simply cannot avoid encountering strong pressure to stop being feminine. Boys who wear dresses or play openly with Barbies are ostracized by at least some of their peers, for example. This means that we can't know how they would grow up if we left them alone. Boys with GID are not left alone.

Imagine that we could create a world in which very feminine boys were not persecuted by other children and their parents allowed them to play however they wanted. Do we really think that boys with GID would have the same low rate of transsexual outcome that they do in our crueler, less tolerant world? As much as I would like to arrange such a world, I think that it might well come with the cost of more transsexual adults.

Maybe it would be worth it, though. It is conceivable to me that transsexuals who avoided the trauma and shame of social ostracism and parental criticism would be happier and better adjusted than the gay men whose masculinity came at the expense of shame and disappointment. Certainly their childhoods and adolescences would be. Perhaps it would be more humane if we educated boys with GID early on that if they wanted, they could eventually become women. If they still wished to become women when puberty began, we could put them on hormones to prevent their bodies from becoming very masculine, so that they would be more realistic and attractive women once they made the change. At age 16, boys who had retained their cross-gender wishes could opt for surgery. I can imagine that this world would be more humane than ours, although we cannot know it without conducting an experiment that will probably never be possible.

In our world very feminine boys must contend with peers who despise sissies, fathers who get squeamish seeing them pick up a doll, parents who have a difficult enough time accepting that their sons will be gay, much less that they might become women. For the most part, people do not just keep these attitudes to themselves but convey them to the boys. So even the boys with GID whose parents don't bring them to therapy are getting at least some therapeutic components. They are getting a regimen of behavioral modification, heavy on punishment. Compared with this, Zucker's therapy seems kinder and more consistent, and thus more likely to be effective. Zucker believes that it is, although he is the first to acknowledge that no scientific studies currently support the effectiveness of what he does. Designing a study that would decide whether his therapy works, over and above the social influence that all feminine boys are guaranteed, is conceptually simple: Randomly assign boys with GID (along with their families) either to receive Zucker's therapy or to receive no therapy at all. See if those Zucker treats are less likely to become transsexual. Or see if the boys Zucker sees are happier in some other way. These are the types of questions that Danny's mother most wanted to know the answers to the day she came to my office. But I could not tell her, because no one knows. Furthermore, given the squeamishness of funding agencies about these kinds of questions, I doubt that we will know the answer for decades, if ever. Which means that parents of very feminine boys are sentenced to acting in ignorance, trusting their instincts, hoping their decisions turn out for the best. Although this is similar to the situation of all parents much of the time, the stakes seem higher for the GID boys.

★★★★★★★★

I am fairly certain that when he grows up, Danny Ryan will become a man rather than changing into a woman. I am more certain that no matter what Danny becomes, his sexual desires will be for men. Now eight years old, Danny probably has not yet had clear sexual desires. Recall that at age five he claimed to dislike boys—he meant that he didn't like their personalities and activities, not that he disliked them sexually. Certainly at age five, Danny had no unambiguous sexual feelings. But he will.

We know very little about how children's sexual feelings develop. Our society is very squeamish about children's sexuality. I am not sure that a study proposing to ask children about their sexual knowledge and feelings could even be conducted in this country at the beginning of the twenty-first century. This would be true especially of a study that aimed to ask about homosexuality. I cannot imagine Congress approving funding for such a project, and many parents wouldn't let their children participate. That is too bad, both for science and for boys like Danny.

Try to remember how ignorant of sex you once were. Well into grade school, I had no idea what a vagina was (despite having two sisters). I thought that intercourse involved the penis going into the anus, or perhaps the navel, or perhaps that sperm crawled from the penis into the woman while people slept—I believed all these things at one time or another. I learned more accurate information gradually, and mostly from peers and experience. But I could count on many peers knowing more than I did, because they were nearly all heterosexual. And regardless of my knowledge of sexual anatomy, I knew that men and women, and many boys and girls, had romantic relationships. I saw evidence for this everywhere: at school, on television, in the movies. My friends and I talked about girls we liked (or pretended not to like).

I didn't know about homosexuality until after grade school, perhaps just before high school, and I had very little idea what it involved. Looking back, I see now that a boy I sat next to and befriended during high school French class was flaming, but I didn't know that at the time. I may have been a slow learner, but my point is that for straight kids, only the graphic details are kept from them—and they have many opportunities to learn these from friends. By comparison, gay kids must feel like Martians. Until very recently, there were no openly gay characters on television or in the movies. Even today, when children hear about homosexuality, it is usually in a derisive way. The gay humorist, David Sedaris, wrote about how important it was for gay children to join straight kids in picking on anyone accused of being gay, in order to direct attention away from themselves. Although there are probably some liberal communities where this would no longer happen, there are many more where anti-gay sentiments are virulent.

Many boys must simultaneously learn that they are gay and that they are despised.

How will Danny learn that he likes men? Commonly, gay men remember that they felt vaguely different from other children. This difference doubtless has something to do with gender nonconformity, but it probably also has to do with sexuality. Even before I knew the correct details about sex, I had crushes on girls. Gay boys presumably have crushes on other boys, and these crushes make them behave and feel different from other boys. Most people recall that they had their first sexual attraction at about age 10, or fifth grade.

A couple of years later genital arousal kicks in, so that boys cannot easily hide their sexual preferences from themselves. Their penises insist on being heard. This is sexual desire. Even here, though, motivated boys can fool themselves. A gay friend told me that he always fantasized about a man and a woman having sex, often with accompanying pornography. He thought this meant he was straight. When he finally admitted to himself that he was gay, he was able to see that he had always been aroused by the men in the fantasies, not the women.

One's first sexual experience is variable in timing, because it depends so much on circumstances. A gay adolescent in a small, conservative community might have no potential sexual outlet. If he is in a large, urban setting, he almost certainly will. On average, gay men have their first homosexual experience at about age 14.

Very feminine gay boys tend to know they are gay earlier than masculine gay boys do. They have been called "gay," "fag," "queer," "homo," and so on, since before they knew the meanings of such words. They are "outed" at an early, pre-sexual, age. When they start having erections around attractive males during puberty, feminine boys need only connect some close dots. In some ways, it might be easier for feminine boys to accept their homosexuality. For example, they do not have to worry about ruining their image. Their image is already gay.

They might also have sex earlier. This is partly because they are quicker to acknowledge their homosexual desire, but it might also be because they are easier for other gay people to recognize. A gay male must be careful about approaching other males sexually, but very feminine boys are a safer bet. I would wager that among the many highly publicized cases of predatory men having sex with adolescent boys, a non-trivial percentage of the boys were recognizably feminine. The older men had reason to think that their advances would succeed.

Early awareness of homosexuality is not necessarily beneficial. Gay men who were gender nonconforming boys and who came out early are more likely to say that they contemplated or attempted suicide than masculine gay men who came out later. We don't know why, but it seems likely to have something to do with the stigmatization of gender nonconformity.

If any feminine boy is likely to have an easy time coming out, it is Danny Ryan. His mother already knows he will probably be gay—I told her—and she says that this won't be a problem for her. She will have to run some interference with her husband, who is much less accepting of the possibility, but she has already learned to do that regarding Danny's feminine behavior. It is odd for me to think that many people would think that Leslie Ryan is shirking her maternal duty by helping Danny feel okay to be gay. I think he is blessed to have her.

★★★★★★★★★

Leslie Ryan says that Danny is "going into the closet more." She doesn't mean the literal closet where he used to seek her shoes. She means that more and more, he is hiding his femininity. Patrick has taken to playing catch with Danny, and Danny apparently enjoys spending this time with his father. But he is not very good at playing catch, and his mother thinks he would rather be doing something else.

He will no longer talk willingly about his feminine ways. Jennifer, his old babysitter, recently visited him. She recalled playing Barbie with him, and Danny said: "We don't talk about those things any more." He seems ashamed to have others know or talk about his un-usual behavior.

He continues to see a therapist, and his mother worries somewhat less about him than she used to. She thinks he has accepted that he will grow up to be a man, if a feminine man. She knows that there are problems ahead too. If Danny becomes a gay man, as seems likely, he will encounter more intolerance. Still, she thinks that at age eight, Danny has left his most difficult times behind him.

CHAPTER 3

The Boy Who Would Not Be a Girl

*I*t is difficult to see a boy like Danny Ryan without wondering why he exists. He is so unusual, and there is no obvious explanation. When Mrs. Ryan came to my office, her main concern was Danny's well being, but even she expressed profound curiosity about where Danny's femininity came from. Of course, her curiosity is tinged with the possibility of guilt. Mothers are the first to be accused of causing their children's problems.

Most people Mrs. Ryan has encountered have probably looked at her first as the likely cause. American views on gender development have been dominated for decades by the idea that differences between boys and girls are rooted in socialization, and socialization begins at home. If a boy thinks he is a girl, the parents must have done something wrong. This has been the commonest view of both scientists, such as developmental psychologists, and laypeople.

The prevailing view is probably incorrect. Danny's atypical behavior is best explained as the result of feelings that began within him and persisted with little encouragement. Recent dramatic scientific findings have suggested that we couldn't socialize a boy to behave like Danny even if we tried.

★★★★★★★★

When she was six months pregnant, Jessica Johnson got news that every expectant mother dreads: something was wrong with her baby. It was not growing fast enough, and when the doctors did an ultrasound, they saw a "mass" on the baby's stomach. They told her that the baby's intestines were at least partially outside its body; but when it was born, this problem could be surgically repaired. Although the consulting surgeon knew Jessica was worried, he reassured her that soon after the baby was born, it would be all over, and she would "never look back."

Mrs. Johnson saw the surgeon again soon after the baby was delivered by cesarean section. He had lost his reassuring demeanor and told her grimly: "Your son's problems are much worse than we expected. He has a very rare and serious problem, called 'cloacal exstrophy.' His bowels are poorly formed and open into his bladder. He needs surgery right now. If they survive, kids like this will need several major surgeries during childhood."

"*If* they survive?" Mrs. Johnson said to herself, but before she could ask him about this, the surgeon added: "There is something else you need to know. Boys with this condition are born with poorly formed penises, and in order to give them the best possible outcome later in life, the standards of care are to surgically reassign them as girls. This means that right away, you need to start thinking of your baby as a girl." He went on to explain that her baby would be castrated and eventually would be given female hormones and have a surgically constructed vagina. She looked at her husband, who seemed dazed, and she burst into tears.

When they talked privately later, they decided that they could live with changing the baby into a girl. They had been planning to call him Jason, but now they agreed to call her Amanda. They would tell no one, and because they had never known Jason, they would not miss him. Although the sex reassignment had been shocking initially, it seemed less important now, in the scheme of things. Amanda had surgery to close her bladder and to correct several other severe problems with her digestive tract. An ileostomy was created, in order for waste to drain. The testes were removed. During the first week, the doctors became more optimistic that Amanda would live.

Her mother tried to find out as much as she could about cloacal exstrophy, its causes and its prognosis. It is very rare, occurring once in every 400,000 live births. The surgeon had seen only one other case. No one knows what causes it. It isn't hereditary, and so if Mrs. Johnson has another child, it will almost certainly not have it. Until the early 1970s babies born with cloacal exstrophy died shortly after birth. When she learned this, Mrs. Johnson briefly wondered whether this would have been a better outcome, but she banished this thought. Children with cloacal exstrophy spend a good part of their childhoods in the hospital, having and recovering from surgeries. And if she and her husband treated Amanda like a normal girl, there was every reason to expect her to become one.

They brought Amanda home when she was a month old. She was their first child, so they had nothing to compare her to, but she seemed to be a happy, easy baby. Aside from the countless trips to the doctor, that is. Mrs. Johnson grew increasingly attached to her.

By age two Amanda had begun to develop a strong personality. She was assertive, loud, and active. When she played with her dolls and cuddly toys, she did so in a very non-maternal way. She would throw them, tear them apart, and carry them by their hair. She enjoyed cartoons meant for boys. By age four, Amanda had begun to balk at being dressed in dresses or in anything pink. She played more with the boys she met than with the girls. On her fourth birthday she had a party, and the mother of one of Amanda's friends said "I think that girl was meant to be a boy." She meant it jokingly, but Mr. and Mrs. Johnson glanced at each other knowingly. Later, for the first time, her husband raised the question that she had not dared to speak: "Do you think we did the right thing?" During their discussion they agreed that Amanda might be a tomboy, but she was a happy child. Neither of them wished her to be different than she was.

Mrs. Johnson brought up the subject of Amanda's behavior with the surgeon the next time she saw him. He told her that it is common for girls like Amanda—that is, girls who were born boys—to have tomboyish traits. "Before she was born, Amanda got a male dose of testosterone, and this is going to have some effect on her brain." But he reassured her that girls like Amanda appear to adjust fine as girls. Mrs. Johnson wondered how certain he could be of this if he had only seen one other patient with cloacal exstrophy, but she didn't want to appear difficult. The surgeon ended their meeting with the warning that it was important that the parents "stick to the plan" of rearing Amanda unambiguously as a girl. "Life is hard enough for her. You don't want to confuse her."

When Amanda entered school, she became even more masculine. During recess she played with the boys, and she began to play sports. She especially liked baseball and basketball. She was a good athlete, as good as the better boys. With her short hair and masculine ways of moving and talking, strangers often mistook her for a boy. Her classmates accepted and liked her, but they recognized that she was different.

When she was seven years old, Amanda's parents began to worry that it was wrong to hide the truth of her birth from her. They felt guilty, and they also worried that when she became an adult she would have access to her medical records and would discover the truth. They worried that she would hate them for not telling her, or worse, that she would not be able to cope with the discovery and would have a breakdown. They decided to tell her.

Although it might seem to be a heavy load to drop on a child, this was no ordinary seven year old. They reasoned that she was probably already wondering why she was so different. Besides, children younger than Amanda learn all kinds of things that others might find disturbing. Mrs. Johnson knew of young children who discovered that they were adopted or who learned that their parents were criminals. Amanda was basically a happy kid, with two loving and supportive parents, and she would be okay.

Amanda's parents told her that they had a "big secret" that they thought she was finally old enough to know. She was, understandably, intrigued. Her mother spoke: "Amanda, when you were born, you were a boy. But the cloacal exstrophy had made it so that you didn't have a normal penis. So the doctors said that you had to become a girl. They operated on you to make you a girl, and they told us that we should raise you as one. I know that is a lot for you to take in, so I'm going to stop and let you think and ask any questions you want."

Amanda sat, still, stunned, expressionless. As her mother and father watched her, they ached inside. She asked: "If I was born a boy, then why can't I still be a boy?" They looked at each other, hesitatingly. She said, more insistently: "I *am* still a boy." Her parents neither agreed nor argued. She ran to her room.

Her parents discussed what they would do. Neither felt certain what the best course of action was. Both were suddenly feeling very disenchanted with the strategy prescribed by Amanda's doctors, to insist that she is a normal girl. Before they could decide a course of action, she emerged from her room and told them that she wanted a boy's name. Mrs. Johnson shrugged and said "When you were born, we were going to call you Jason." Amanda/Jason said: "I like that name."

It was July, and school was out, so Mr. and Mrs. Johnson felt that they had a bit of time to reach a resolution. Their child was determined to stay a boy, and when they forgot and called her "Amanda," she corrected them impatiently: "Jason!" They were surprised at how easily they stopped making that mistake. By August, they were firmly on Jason's side. Jason brought up the question of how they would let his classmates and friends know about his change, and he decided that they would tell everyone on the first day of school. They told his teacher beforehand. (When they told her they had a serious issue to discuss, she insisted that she had heard everything before. When they told her, she changed her mind.) It was a small class, and parents were told ahead of time that an important announcement would be made. So the room was full of curious parents and children on the first day.

Mr. Johnson began: "Many of you know that our child has always seemed more like a boy than a girl. Well, actually, Amanda was born a boy, and changed into a girl due to medical problems. But Amanda has decided to be a boy after all, and his new name is 'Jason.' Any questions?" The class sat in awkward silence until one parent said: "Jason, what's your favorite sport?"

The children seemed to adjust quickly. It was as if people change sex all the time. A few days later Jason said, "The day I became a boy was the happiest day of my life." He has said that many times since. He is the best player on his junior high school basketball team, and he has a girlfriend. His parents say that it is difficult to imagine him as Amanda. They have no doubts that they did the right thing.

★★★★★★★★

Suppose we wanted to do the perfect experiment to determine whether the essence of boyhood and girlhood is inborn or learned, the perfect nature-nurture study of gender. What would we do?

Just after conception, male and female fetuses are quite similar. What make them differ are the direct and indirect effects of testosterone, which is present in much higher levels in males. This is why, for example, males develop penises and females clitorises. Many scientists believe that there are important brain differences between newborn boys and girls that contribute to later behavioral differences. Other scientists believe that at birth the brains of boys and girls are essentially identical, and that girls and boys behave differently entirely due to the socialization they receive.

So to the perfect experiment: First we would take normal newborn boys from their mothers. We would castrate the boys and surgically give the babies vaginas. Next, we would give them away to unsuspecting parents, whom we would lead to believe were adopting girls. We would watch the children to see how they develop.

What would we watch for? One obvious thing would be to see whether the children behave more like boys or girls: whether they had more stereotypically masculine or feminine interests; whether they played rough; whether they had boys or girls as friends; whether they liked wearing frills and having long hair or preferred pants and short hair. The degree to which children behave like stereotypical boys or girls is sometimes called *gender role behavior*. A second important outcome to assess, when the kids got old enough, would be *sexual orientation*. Do the adolescent boys-changed-to-girls develop crushes on boys or girls? We would also want to know whether the children were satisfied being girls, or whether they would prefer returning to their male state. This depends on *gender identity*, which we have already encountered.

Unfortunately for scientific progress—but fortunately for those of us who prefer humane societies—the definitive experiment can never be performed. It can, however, be approximated, due to misfortune. For example, science and the media have given a great deal of attention to two cases of boys who lost their penises in infancy and who were reared as girls.

The first, more famous, case is that of David Reimer. (Before he publicly revealed his name, the case was known as the John/Joan case.) As an eight-month-old baby, David lost his penis in a surgical accident. The prevailing scientific belief at the time was that children were "psychosexually neutral at birth." That is, both boys and girls could be changed into the other sex, with the right upbringing and the right surgery. However, it was, and still is, beyond surgeons' abilities to construct penises that function acceptably. On the other hand, surgeons have become expert at constructing vaginas. And so David's parents were given the choice of whether to raise him as a boy without a penis or a girl with a functioning vagina. With medical advice, they decided to raise David as a girl. This case was especially exotic because Reimer has an identical twin brother who was raised alongside David as a normal boy.

Early scientific reports asserted that this boy-turned-girl was functioning well as a slightly tomboyish girl. However, we know now that David Reimer experienced a great deal of inner and external torment while being reared as a girl. Other children called her "Cavewoman," because she moved in such an ungraceful manner. She disliked the feminine accoutrements forced on her by her mother. After receiving estrogen therapy at age 12 she grew breasts, but this only contrasted more with her hypermasculine appearance and made her feel more freakish. When she was 14, she became completely fed up and stopped trying to conform to a feminine stereotype. For example, she began to urinate standing up. This caused great friction with her female high school classmates, who stopped allowing her to use the girls' bathroom. She also refused to cooperate with further medical efforts to feminize her. One of the physicians involved advised her parents that it was time to tell her about her past. When they did, she felt stunned, but oddly justified: "Suddenly it all made sense why I felt the way I did. I wasn't some sort of weirdo." Immediately, she decided to change back into a male, and became David again. Although he had some difficulties adjusting (sexual relationships were initially embarrassing and difficult), he never regretted his decision, and is currently married to a woman.

The Reimer case diverged from the perfect experiment in at least one important respect. Reimer lost his penis at age 8 months, and it was not until he was 17 months old that the family decided to rear him as a girl. Although David does not remember anything from that time, it is very likely that he had already begun learning, including about his sex. Perhaps he was simply too old to make the transition. A second very similar case, with sex reassignment before age 6 months, appears to have had a different outcome. In this case, the boy-who-became-a-girl has stayed that way. (On the other hand, she is a lesbian with very masculine interests.) With just these two cases, the position that gender identity, at least, is a matter of very early upbringing is defensible.

Cloacal exstrophy, Jason Johnson's condition, is even closer to the perfect experiment than surgical accidents. There are three main reasons. First, for 20 years or so, most boys born with cloacal exstrophy were castrated and reassigned to girls within days after birth. Second, cloacal exstrophy is such a serious condition, with so many different medical consequences, that parents are less likely to obsess about the sex reassignment than they would be if that were the only problem a child had to face. So it is less plausible that the parents of a cloacal boy-to-girl would have a problem seeing and raising their child as a girl. The third reason is that cloacal exstrophy, although very rare, is much more common than accidents (penile ablation) that cause infant boys to be reassigned as girls. There are only two cases of penile ablation that are reasonably well documented in the scientific literature. Although the results of these two cases are fascinating, they are too easily dismissed as "just two cases."

In contrast, one scientist has been studying boys born with cloacal exstrophy who have (mostly) been reassigned as girls, and he has followed a number of them into childhood, and some beyond. His results are likely to provoke a revolution in the science of gender identity.

****

Johns Hopkins University in Baltimore, Maryland, is one of the world's greatest academic hospitals. This is where psychologist John Money did most of his work on the development of gender identity. Money was one of the most important scientists of the twentieth century, and his work on "pseudo-hermaphrodites"—people whose biological sex is neither clearly male nor clearly female—led him to believe in psychosexual neutrality at birth. Money is the intellectual father of reassigning boys with damaged penises as girls, provided it is done early enough. In fact, Money advised David Reimer's parents to reassign him a girl. Reimer's anguish as a girl and return to the male role, and the case's publicity, have seriously damaged Money's reputation, but this does not diminish his importance in the history of the science of sex and gender.

Like Money, William Reiner is at Hopkins. This is an ironic coincidence, because his work threatens to undermine Money's theory of psychosexual neutrality at birth. Reiner originally trained as a urologist, and for 12 years was a practicing surgeon. During this time he gained experience reconstructing anomalous genitalia of children and adolescents with conditions including cloacal exstrophy. He became increasingly fascinated by the psychological development of these children, so much so that he retrained in psychiatry. Says Reiner: "I no longer wished to 'fix' children's genital abnormalities—I wanted to find what makes them grow, mature, and figure out who they are, regardless of their genital realities. And anyway, the pre-op and post-op anxiety was killing me." He began studying the psychosexual development of children with cloacal exstrophy in 1993. Today he heads the Gender Identity and Psychosexual Disorders Clinic (Child and Adolescent) at Johns Hopkins. His style is direct and passionate, and along with his groundbreaking scientific study, this has made him a controversial figure.

Reiner recently submitted a major scientific article on the outcome of boys with cloacal exstrophy reassigned at birth as girls. Most of the children were teenagers (ages 14-20) at last follow-up. Of 14 female-assigned children, 7 have declared that they are boys. Five of these 7 did this spontaneously. For example, one child refused at age 12 to begin estrogen therapy, saying "I am a boy." In another case, the child was hospitalized for depression before declaring that she was male and wanted a penis. In the non-spontaneous cases, the change occurred as it did with Jason, after the parents came clean about the child's birth.

In two cases in which the children spontaneously declared they were boys, the parents refused to acquiesce to the child's wishes to change sex. These children remain girls to their parents, but maintain male identities elsewhere. This will presumably change when they grow up and assume complete control of their identities.

What about the children who maintain their female identities? One had wished to become a boy but accepted her status as a girl. Later, her parents told her about her past, and she became angry and withdrawn, refusing to discuss the matter. Parents of the others are determined that the girls will never find out about their birth status. Three have become withdrawn, and a fourth has no friends.

Two other children that Reiner has followed were reared as boys because their parents refused sex reassignment. (Not all parents had this choice. One of the parents I spoke with was threatened with child protective services if he refused to allow his child to be reassigned.) Both of these boys are happy, typically masculine boys, although one is concerned about his sex life without a normal penis.

All Reiner's cases who have talked about sexual and romantic feelings are attracted to females; however, several have not revealed anything about these feelings. All cases have unfeminine interests and behavior, and some are quite masculine. One who returned to the male role has been arrested for assault.

What do we make of these results? Just looking at the numbers— 7 of 14 reassigned children returning to the male role, 7 of 14 remaining female—one might be tempted to conclude that no generalization is possible. But it is very rare for a girl to renounce her biological sex in the insistent way that the first 7 did. The 50 percent rate so far among Reiner's kids is extraordinarily high.

And even though half the children remain girls, what is our best guess about their state of mind? Do they represent successful adjustment to the female role? Let's assume that none of these children will renounce her status as a female and that in adulthood all will consider themselves women. Does this mean that they have normal female gender identity? Will they be happier as women with functioning vaginas than they would be as men with non-functioning (or absent) penises?

In their pursuit of the perfect nature-nurture experiment, scientists have thought too little about how to assess outcome. Cleary, if a male infant is reassigned as a girl and later declares that he is a boy, the initial reassignment was a mistake. But if the child does not openly renounce the female role, does this mean that the decision to reassign to the female role was correct? For the most part, scientists such as John Money have acted as if this was a correct inference. But scientists have not fully appreciated how complicated a trait gender identity likely is, or how little we know about it. One expert told me, bluntly: "Gender identity is defined as 'the inner sense of oneself as male or female.' What the hell does that mean?"

One under-appreciated complication is that gender identity is probably not a binary, black-and-white characteristic. Scientists continue to measure gender identity as "male" or "female," despite the fact that there are undoubtedly gradations in inner experience between the girl who loves pink frilly dresses and cannot imagine becoming a boy and the extremely masculine boy who shudders to think of becoming a girl.

A second complication is the translation of inner experience to words. Of course scientists recognize that sometimes people don't reveal everything that is on their minds, and so a cloacal exstrophy child might not openly admit the preference to become a boy. But how would a girl even know if she had the same inner experience as a typical boy? If she had been reared from birth as a girl and had no notion that sometimes boys become girls and vice versa, would she still have the conscious realization that she was a boy inside? I think that the answer to this latter question is "quite possibly, no." If I am right, then scientists have been using a very biased definition of errors in gender identity. The bias is toward missing mistakes.

The perfect nature-nurture experiment requires a better way of measuring the outcome than merely waiting to see if a child spontaneously asks for a sex change. What we really want to know is whether a particular child would be happier being reared as a male or as a female. Of course, no one can go back in time, and so we can't get a complete replay of anyone's life as both a male and a female. However, Reiner's results all point to the superiority of male assignment for cloacal exstrophy cases born male. This is obvious for those who changed back to boys. I spoke to parents of three of these children, and all said their children were much happier as boys than they had been as girls. Interestingly, only one of these parents said her child had seemed unhappy as a girl. The other two characterized their children as basically happy before and yet much happier after becoming boys. Both of these cases were non-spontaneous changes—the children changed back to boys after their parents told them about their childhoods. It is certainly possible that these children would have stayed girls without their parents' revelation. If this happened, scientists who studied them would probably say that they had successfully adjusted to the female role. But this would have been misleading in a serious way.

The children who remain girls are especially poignant. With one exception, none knows her birth status. Reiner's descriptions of them suggest that they are less than happy. Indeed, Reiner thinks that all the cloacal cases born as boys would be happier as boys rather than girls, because their brains have been biologically prepared for the male role. He thinks that those who remain girls are at best missing out, and at worst are experiencing great inner torment. He thinks their parents should tell them and, essentially, let them choose their sex.

Reiner's results, on top of the publicity surrounding the David Reimer case, have provoked a reconsideration of the practices and beliefs of the past 25 years. This is not to say that Reiner has persuaded everyone. Reiner's results are so contrary to expectations that some scientists have privately questioned them. The skeptics don't think that Reiner is making up his results. Rather, their complaint is that he must be doing something to cause the high rate of gender identity change in the cloacal exstrophy children he has studied. Perhaps he asks leading questions that cause children to question their gender identity more than they otherwise would. Perhaps he encourages parents to reconsider the wisdom of rearing the children as girls. (However, the parents I spoke to denied that Reiner did this.) I find it difficult to imagine anything Reiner might have done that could have been so extreme as to make an otherwise happily adjusted child want to reverse sex. I think of my own daughter and cannot imagine her deciding to be a boy, even if I lied to her and told her that she was born one. If Reiner has gone further than other researchers—and at this point there is no evidence that he has—it has had the result of providing a more accurate scientific picture and a more humane outcome for boys born with cloacal exstrophy.

★★★★★★★★★

I imagine introducing the Ryan and Johnson families to each other. I would like Mrs. Ryan to meet Jason Johnson, to hear about his history, to see him now. I'd like to see Danny and Jason together. I'd like to show them to a stranger and ask "Which of these boys do you think was raised as a girl?"

Jason Johnson was castrated at birth, told he was a girl named Amanda, and at least initially, treated like a girl. But it didn't take, and now he is Jason again. Danny Ryan was raised as a boy, and at times harshly punished for not acting like one. He has arguably become more boy-like, at least on the surface. But where did Danny's extreme femininity come from in the first place?

Theories about boys like Danny range from nurture to nature. The nurture hypotheses include the idea that Danny's mother consciously or unconsciously wanted a girl and so undermined his masculine development; that Mrs. Ryan was so unavailable to Danny during infancy that Danny became his mother, in effect, in order to always have her with him; and that more generally, Danny's parents' socialization of Danny as a boy was inconsistent and ambivalent.

I find these ideas to be implausible, and I have named their ilk "Looked at 'em funny" theories of gender identity. According to these theories, you can call a child a girl (or boy), give her (or him) a sex-typical name and the stereotypical toys and clothing associated with her (his) sex, but what really matters are very subtle features of parent-child interaction. These subtle features are not usually well specified, and when they are, they don't appear to be that unusual or specifically linked to gender identity problems. For example, Mrs. Ryan was depressed and inattentive when Danny was born, but most depressed, inattentive mothers don't have sons like Danny. If gender identity development could be undermined by subtle miscommunication between parents and their children, then gender identity problems would not be rare. But they are rare, except in cases of boys castrated at birth and reared as girls.

What about nature theories? In general, the theory that Danny's femininity is inborn would begin with the idea that prior to birth his brain was not masculinized the same way that Jason Johnson's was. Because male hormones such as testosterone are probably responsible for making boys' brains masculine, we would infer that Danny's brain was either exposed to low levels of testosterone or insensitive to testosterone's effects.

The one difficulty with this hypothesis is that anatomically, Danny appears to be a normal boy. If Danny's body had little or no testosterone during all of prenatal life, it would show. Danny wouldn't have a normal penis, for example. It is possible that hormonal effects on the brain occur after penile development. Classic research on monkeys shows how this might work. When female rhesus monkeys were given testosterone in the womb, effects depended on the timing. If they received the hormone during their first trimester they had enlarged clitorises but played like female monkeys. If they received only a late dose, they had normal female genitalia but played rough, like male monkeys. Perhaps Danny's testosterone levels were normal during his early prenatal life but low, for a boy, during later prenatal life.

The fact is that we don't know enough about hormonal effects on the human brain to have a very specific theory of how Danny's brain could have developed in a feminine direction while his body developed masculine. If Danny's body also showed signs of feminine development, this would support nature theory, but the lack of anatomical femininity does not disprove it.

There has been essentially no research on boys like Danny that is directly biological. Short of dissecting the brain of a feminine boy and comparing it with normal boys' and girls' brains, it is unclear what we would even look for. However, the best conceivable direct test of nurture theory has been tried, and it failed. Amanda became Jason again.

Part II

The Man
He Might Become

*T*he six men addressing my under-graduate sexuality class have two things in common. First, they all look fabulous: fit, muscular men, with square jaws, short neat hair, and stylish masculine clothes. They look like models from J. Crew or Ba-nana Republic catalogues, which may be one reason why many more female than male students are asking them questions. I see the looks on the women's faces as they listen to the panel, and they convey wistful attraction. This is due to the hopeless nature of the attraction—hopeless not because the men are 10 years older than my students, but because of the second thing the men have in common: they are all gay.

Because the class's subject is sexuality, I have asked my students not to hold back from asking questions of interest even if the questions are personal or explicit. (The men on the panel have assured me that such questions are okay.) The students eagerly oblige.

"How and when did you come out to your family?" Answers ranged from Rick's "I haven't yet" to Ben's humorous account of telling his mother: "She was visiting me at college and I took her out to dinner. I told her 'I have something to tell you,' and she looked very worried. At that point the waiter leaned over and said to me 'Just tell her honey!' When I told her, she was relieved and said that she had been afraid I was angry at her."

"Did you ever have sex with a woman?" Four of the guys have (two enjoyed it, and two did not), and two have not.

"Can you give the girls in the class some oral sex tips?" The men agreed that it is important to actually enjoy giving oral sex, and not to use one's teeth.

"Do you really enjoy it when a man with a large penis has anal sex with you?" Answer: "Honey, you don't know what you're missing."

"Professor Bailey says that gay men are usually feminine during childhood. Does that describe your childhoods?" I am happy that someone has brought this up, and I am eager to hear the panel's responses. Ben says, "I wasn't much different than other boys. What about the rest of you guys? Anyone want to say anything?" For a few moments the remaining men look at each other and shrug, and then Ben says "Next question?"

I am disappointed with the lost opportunity to hear recollections of childhood femininity. To be sure, many gay men do not recall being markedly feminine boys, and a few even recall being more masculine than average. But I suspect that this panel does not consist only of gay men with masculine boyhoods. Rather, I think the guys avoided the question. This explanation is consistent with their body language and their eagerness to go on to the next question. It is also consistent with my past experience talking with many gay men about femininity, especially femininity during childhood.

I immediately think of two episodes during my career as a scientist studying this issue. The earliest occurred in Dallas, where I had traveled to interview gay twins for a study regarding the genetics of sexual orientation. I had a standard interview, which included questions about childhood gender nonconformity. ("Were you ever called a sissy?" "Did you ever dress up in girls' clothes?" and so on.) I had noticed that during this part of the interview some of the gay twins looked uncomfortable. One twin in Dallas took a long time to answer—he had, in fact, been a very feminine boy—and then he told me, "I haven't thought about those things in years." I think he wished I hadn't made him remember.

The second incident occurred recently when I gave a talk at a conference on sexual orientation. During my talk I showed a short video of a feminine boy dressing in girls' clothes and playing with dolls. Afterwards, a local gay politician approached me, smiling uncomfortably. He thanked me for my presentation and said that he thought it was extremely important work. But he confessed that watching the boy in the video was a wrenching, "obscene" experience for him. He had just revisited his own childhood from his present perspective and found it disturbing.

Reactions like these have been common among the gay men I've spoken to about childhood femininity. In fact, of all the controversial topics related to male homosexuality, the contention that gay men tend to have been feminine boys (and may be feminine men) has provoked the most discomfort and dispute. Initially, I found this odd, because the link between childhood gender nonconformity and adult homosexuality is one of the largest and best established associations regarding sexual orientation. But after repeatedly encountering this kind of reaction, I began to think something interesting was going on. I made up a word to describe gay men's attitude: femiphobia. (Independently, the writer Tim Bergling came up with "sissyphobia.")

Why are gay men femiphobic? Part of it is adverse childhood experience. I don't think that either the gay twin or the gay politician would endorse the belief that childhood femininity is a bad thing, but both behaved as if it were something to be ashamed of. I inferred that as boys, both men had been subject to the shame-inducing disapproval of others, including parents and peers. To be reminded of this is unsettling. But I have come to realize that it is not only childhood mistreatment that causes gay men to react negatively to the suggestion that they are, or were, feminine. To explain the other reasons requires some additional knowledge, and so I will return to them.

✶✶✶✶✶✶✶✶

I live in a section of Chicago called "Wrigleyville," due to its proximity to Wrigley Field, the home of the Chicago Cubs. I live between two major streets. Half a block to the west is Clark Street, which borders Wrigley Field, and which contains scores of singles bars filled with young heterosexual people. Half a block to the east is Halsted Street, which is the central artery of "Boy's Town," Chicago's historic gay district. Halsted is lined with gay bars, filled mainly with gay men. (My favorite names are "Manhole" and "Cocktail.") I visit with friends in both places, sometimes during the same evening, and it is difficult for me to do so without my scientist hat. On some nights I am struck by the differences between gay and heterosexual men. On other nights I am impressed by their similarity. It all depends on which aspects of behavior I am focusing on.

Gay men are comprised of a mixture of male-typical and female-typical characteristics. The idea that gay and straight men differ only in their preferred sex partners is wrong. The most general way in which they differ is that many gay men are somewhat feminine in certain respects. But even the most feminine gay men are not merely women with penises. There are ways in which gay men are every bit as masculine as heterosexual men, and indeed, may even appear more masculine. Psychologist Sandra Witelson has hypothesized that the brains of homosexual people may be mosaics of male and female parts, and I think she is right. This mixture explains much of what is unique in gay men's culture and lives.

CHAPTER 4

Gay Femininity

anny Ryan (the boy from Part I) will probably grow up to be a gay man, but does this mean that most gay men were boys like Danny Ryan? Not necessarily. Perhaps men who were very feminine boys comprise only a very small subgroup of gay men, the rest of whom had been just like other boys.

The easiest way to address this is to ask gay and straight men about their childhoods. I have by now discussed childhood behavior with hundreds of gay men, and my general impression is that the typical gay man is noticeably more feminine than the typical straight man, but that the degree of femininity that Danny Ryan displayed is rare. (At least few gay men remember or acknowledge that degree of femininity.)

We don't have to rely on my impressions, however, because many objective studies have been done on this question. In these studies, gay and straight men are asked about their childhoods using questions like those below.

Rate your agreement with each item, from 1 (strongly disagree) to 7 (strongly agree).

As a child I was called a "sissy" by my peers.
As a child I sometimes wished I had been born a girl rather than a boy.
As a child I preferred playing with girls rather than boys.
As a child I often felt that I had more in common with girls than boys.
As a child I sometimes wore feminine clothing (such as dresses), makeup, or jewelry.
As a child I disliked competitive sports such as football, baseball, and basketball.
I was a feminine boy.

In 1995, Ken Zucker and I reviewed more than 30 studies that had given questionnaires like this to gay and straight men. We found that, on average, gay men were much more feminine in their memories than straight men. The size of d (the effect size, which shows how large a difference between two groups is) was 1.3, which is considered quite large by conventional scientists. Few interesting findings in the behavioral sciences are this large. We estimated that the typical gay man was more feminine than about 90 percent of straight men. On the questions above, the average straight man gets an average of less than 2 (on the 7-point scale); the average gay man gets about a 4.

Of course, memories can be wrong. If there was some tendency for gay men to exaggerate how feminine they were, or for straight men to understate how feminine they were, these numbers could be off. I have already suggested that gay men tend not to embrace the idea that they are, or were, feminine, so I don't think that gay men are likely to be exaggerating. Moreover, we have both prospective studies like Richard Green's showing that feminine boys become gay men, as well as retrospective studies like ours showing that many gay men were feminine boys. The simplest explanation is that both of these findings are true.

One interesting observation about gay men's memories is that they are noticeably more variable than straight men's. For example, on the items above, straight men tend to be bunched up at the low end of the scale, meaning that they are denying any significant femininity during childhood. In contrast, gay men show more range in their scores. Perhaps 20 percent of gay men have scores similar to the straight men, but a significant minority has scores around 6, meaning that they are agreeing to every feminine item. If we can trust gay men's memories, then some were feminine boys and others masculine. This could mean that there are different types of gay men. At the very least it means that we should be aware of potential variation. For example, gay men who were feminine boys are likely to differ from gay men who were masculine boys in adulthood as well. Anyone who knows a few gay men has noticed that they vary considerably in how gay-stereotypical their behavior is. This variation may well begin in childhood.

****

It is a crisp October Sunday, and I am meeting Ben (the leader of the "gay guys panel" who spoke to my human sexuality class) and some other friends at Sidetracks on Halsted Street. It is about 3:30 PM, and the Chicago Bears are playing. On Clark Street, a block away, countless sports bars have the game on their large-screen projectors, while men wearing Bears jerseys flirt with similarly dressed women. But not at Sidetracks. Here, the largely male crowd is watching video performances of show tunes and singing along with Fred Astaire, John Travolta, and Gordon MacRae. (Instead of "Oklahoma OK!" they sing "I'm a homo, OK!") I imagine switching channels at Sidetracks and some Clark Street sports bar so that the straight guys have to watch Gene Kelley and the gay guys have to watch the Monsters of the Midway, and I find the thought amusing.

Stereotypes about gay men—and about straight men—start with their interests. Gay men enjoy show tunes, acting (more generally, the arts), fashion, decorating, dancing, and lots of sex. Heterosexual men enjoy football, baseball, basketball, hockey, shopping for stereo equipment and cars, and lots of sex. For about 30 years, from the late 1960s until the late 1990s, it was de rigueur to scoff at these stereotypes and look askance at those who believed them. But recently, science has provided support for the stereotypes, in the only way that stereotypes are ever true: on average.

Of course, not all gay men like fashion, and some heterosexual men do. Let me save us a lot of tedious qualification by admitting that not all gay men are alike, and not all straight men are alike, and some gay men are very much like straight men (except, by definition, in their sexual orientation). This is important to keep in mind, but it does not invalidate the fact that there are some large differences between typical gay men and typical straight men.

The leading researcher in this domain has been the psychologist, Richard Lippa. Lippa's contribution has been doubly reactionary, because he has confirmed stereotypes about differences both between men and women and between heterosexual and homosexual people. Lippa devised questionnaire measures of both occupational and recreational interests. ("Rate your interest in being a jet pilot, nurse, fashion designer, physicist….; going to art galleries, surfing, computers, aerobics….") Altogether, more than 100 occupations and hobbies are rated. One can then compute a total score on the questionnaire, that I will call "Feminine Interests," by adding all the ratings for stereotypically feminine interests and subtracting all the ratings for masculine interests. He originally studied male and female college students, who might be expected to be less subject to stereotypes than other people. For example, few of my students, male or female, would say that men and women should be encouraged to pursue different careers. Many of them doubt that their interests diverge from those of the other sex. From my classes at Northwestern University, few women will become nurses, and few men will become jet pilots. Still, even at Northwestern, Lippa's scales yielded huge differences. Conventionally, sex differences (and more generally, differences between two groups) are measured as "*d*": *d* values of 0.2 or less are considered small; *d*s up to 0.8 are considered moderate; and *d*s above 0.8 are considered large. The sex difference in height is about 2.0, which is whopping. One would have to be perverse to deny that the sexes differ in height, on average, although of course (for the last time), some women are taller than some men. Lippa's scale usually yields a *d* between men and women that approximates the size of the sex difference in height.

Stereotypes about gay men (and thus, necessarily, straight men) include occupational and recreational differences. (Quick: One man is a hairdresser, and the other is a Marine sergeant. Which one is gay?) In one sense, then, it is obvious that Lippa would use his questionnaire to study sexual orientation differences. Knowing the sensibility of academic psychology in the late 1990s, however, Lippa's research seems bold. In several studies, he has found that gay men are midway between heterosexual men and women in their sex-typed interests. The sexual orientation differences are large, although only half the sex difference. I have found similar results using Lippa's questionnaire with a non-student sample. Furthermore, the gay men who were most feminine during boyhood tend to have the most feminine interests as adults. Perhaps not surprisingly, feminine occupational interests appear to be the continuation of feminine childhood interests.

There is more than one take on these findings, however. Many people find the idea of intrinsically feminine interests to be preposterous. They think that what men find interesting and what women find interesting are socially arbitrary. Somehow, "society" decides that certain tasks are masculine and others feminine, and because of this decision, men and women are socialized to do different things, in order to conform. Differences, both between men and women and between gay and heterosexual men, reflect historical accidents rather than more fundamental differences. I call the idea that men and women (and gay and straight men) have cross-culturally consistent (and probably innate) differences in interests the "psychological" hypothesis. This is because it suggests that the sexes, and the sexual orientations, are really psychologically different. The competing idea, that these differences are largely arbitrary, is known as the "sociological" hypothesis. This hypothesis implies that men and women, or gay and straight men, are the same psychologically, and that behavioral differences between them reflect sociological factors such as group identity. Obviously, the question is at least somewhat amenable to scientific study. For example, we can see whether the kind of activities preferred by gay men has some consistency across cultures.

But before we study gay men in other cultures, it is useful to know what they are like in ours. Here in Chicago just past the turn of the century, I think I observe a preponderance of gay men in the following occupations: florists, waiters, hair stylists, actors (or at least acting students), classical musicians (but not rock musicians), psychologists (or at least psychology students) and psychiatrists, antique sellers, fashion and interior designers, yoga and aerobics instructors, masseurs, librarians, flight attendants, nurses, clothing retail salesmen (e.g., at the Gap and Banana Republic), web designers (but not software or hardware designers), and Catholic priests. Assuming I'm right—and I may not be in some cases—do these occupations have anything in common? One main thing is that with the exception of the priesthood (from which they are barred), women express higher than average interest in them. Another piece of the picture, noticed by Lippa, is that many (but not all) of these occupations require interacting in a social context. A major distinction between different occupations is that some require interacting with people while others are more focused on inanimate things such as machines. On average, the occupations I listed—especially nursing, retail sales, psychology, waiting tables, and flight attending—are relatively high on the "people" side of this distinction. But that hardly captures everything. There is also an aesthetic/artistic component (acting, designing, and even collecting antiques reflect this). The other distinction I think figures in some of them (clothing retail sales, hair styling, aerobics) is a concern with physical appearance.

Although I am fairly certain that a well-done scientific study would find disproportionate numbers of gay men in the occupations I listed, the definitive, comprehensive study hasn't been done. I have done the only study I know of about a particular occupation. It is an occupation I purposefully omitted from the list above. Did you notice something missing?

****

Around 1995 a Northwestern undergraduate who was interested in a question related to homosexuality approached me. The undergraduate, Michael Oberschneider, was somewhat older, in his late twenties, and appeared to be straight. (Among other indications he mentioned a girlfriend.) He explained that he was delayed en route to college by a career in ballet, including a stint in the Boston Ballet Company. He had long wondered why so many of his fellow male dancers were gay. I had wondered about this, too, and suggested collaborating. Michael agreed, and he embarked on a most ambitious study. Before it was over, he had interviewed 136 professional dancers from around the country, including several well-known choreographers: 48 gay men, 42 heterosexual men, 45 heterosexual women, and 1 lesbian.

We got our participants haphazardly, primarily from Michael's professional connections and their connections, and so on, and therefore, they do not comprise a random or representative sample. Still, we presumed that they knew much more about the sexual orientation of professional dancers in general than other people did. We asked each participant to give an estimate of the percentage of male dancers who are gay. On average, they estimated 58 percent; the smallest percentage anyone gave was 25 percent. Compared with a rate of 2-4 percent in the general population, this is a huge difference. The average proportion of gay men in the dancers' own companies—which presumably they could estimate fairly accurately—was 53 percent.

We interviewed only one lesbian dancer, because she was the only one we could find. Consistent with this, all dancers gave low estimates for the rate of lesbianism among female dancers, for an average of 3 percent. Professional dance is not generally a homosexual occupation; it is a gay male occupation.

We also asked participants how and when they got interested in dancing. The gay men actually got started a couple of years later than straight men (age 13 compared to 11). When asked what initially motivated them, about 60 percent of heterosexual dancers said that their parents had encouraged them. (Heterosexual men and women gave similar responses.) In contrast, only 13 percent of gay men said they got into dance this way. Instead, half the gay men said they got interested in dance by themselves, compared with only 19 percent of the straight dancers. For example, one gay dancer recalled watching the Jackie Gleason show at age 6 and seeing the June Taylor Dancers. Immediately, he decided that was his career goal. Although he was unable to obtain a position with the June Taylor Dancers (who were all women), he became a prominent choreographer.

We also gave the male dancers questionnaires about memories of childhood femininity. We expected to find smaller sexual orientation differences than usual, because we thought that straight male dancers probably were not the most masculine boys. Instead, we found larger differences than usual. Contrary to our expectations, the straight male dancers were similar to other heterosexual men to whom we have given the questionnaire. The gay male dancers recalled especially feminine childhoods compared with the straight dancers, but also with other samples of gay men we have studied.

Overall, our study suggests that many gay dancers were very feminine boys who discovered dance on their own. Pursuing a career in dance requires a great deal of both talent and dedication. What is it about feminine boys that ensures that as adults, they will be overrepresented in professional dance by about a factor of twenty? It is possible that such boys possess some innate talent that makes them good dancers. Because I do not have an ounce of relevant ability, it is difficult for me to imagine a specific hypothesis. I think that the more significant part of the story must be the intense early interest that feminine boys have. My son was 10 years old when we began our dance study. One day I explained what we were studying, and I asked him why I might expect to find a high rate of gay male dancers. He immediately answered, "Because dancing is feminine, and gay men tend to be feminine." I was pleased by his answer, which was also mine. It seemed to me that if a 10-year-old boy could come up with the hypothesis, then scientific reviewers of our study would not find it far-fetched. (Indeed, some of my friends made fun of me for studying something they already considered to be an obvious fact.)

Danny Ryan (our feminine boy from Chapter 1) has recently begun ballet lessons. He has also been to the opera, and he enjoyed it. (How many eight-year-old boys enjoy the opera?) He has attended mass and has already asked to become an altar boy when he is old enough. (His mother thinks that this has to do with the altar boys' costumes, which look like dresses.) Because of the intense competition for these careers, he will probably not become an opera singer, ballet dancer, or priest. He is more likely to become an accountant. Still, twenty years from now on any October Sunday, he is more likely to be singing show tunes somewhere than to be cheering for the Chicago Bears.

****

The actor from the Second City comedy troupe plays several roles this evening, and one of them is a gay man. We know almost instantly when the actor becomes that character—not because he says "I'm gay" or puts the moves on any men or mentions a feminine occupation. It's the way he talks. Male comedians in the United States are often adept at affecting, for wont of a better term, a "gay accent." This suggests two things. First, people recognize the accent, and so perhaps some gay men really do speak that way. Second, people must think that a character with a gay accent is funny.

Another anecdote involving my son: When he was 10 years old, we were sitting in a theater waiting for the movie to start. A man behind us was speaking, and my son leaned over and said, "Dad, there's someone for you to study." My son knows that I study sexual orientation, and this was his way of suggesting that the man sounded gay. The content of the man's speech was unremarkable, and so the only clue my son had was the way the man was speaking. Immediately before my son made his remark, I had had the same intuition. I was struck that a 10-year-old boy could have absorbed this cultural stereotype— I had never talked with him about it. Of course, I did not get the opportunity to check the accuracy of our judgments by asking the man about his sexuality.

I cannot imitate the gay accent, and I cannot even describe it, but chances are, you know what I'm talking about. Done well, it does not include a lisp, which is the lazy straight man's way of pretending to speak like a gay man. Before worrying much what the gay accent is, it seems more important to determine whether there is any truth in the notion that gay men speak differently. And so I did the following study.

We recruited homosexual and heterosexual men and women to the lab to provide several types of data. We got approximately 30 from each group; the relevant groups here are gay and heterosexual men. The relevant data, for now, consisted of short speech samples. Every subject read the Harvard Sentences, a collection of sentences that are interesting to linguists because they contain all the phonemes (elemental sounds) of the English language. Some example sentences include:

It's easy to tell the depth of a well.
Four hours of steady work faced us.
Help the woman get back to her feet.
The soft cushion broke the man's fall.

Subjects read the sentences into a microphone connected to a computer, which stored the recordings. Next, we recruited an entirely different sample of homosexual and heterosexual men and women to listen to the four sentences given above. This new sample (Listeners) rated each person in the first sample (Speakers) on a scale from 1 (very heterosexual sounding) to 7 (very gay sounding).

Results were striking. The size of *d* (the effect size) was about 2.0, as large a difference between gay and heterosexual men as I have observed (except, of course, when we ask about their sexual orientations). Only 10 percent of the heterosexual men were rated above 4, on average—4 represents a "neutral" score of neither gay nor heterosexual. In contrast, 75 percent of the gay men were rated above 4. Excluding one very unusual straight speaker (rated a 4.9), two-thirds of gay speakers were rated as gayer sounding than any straight speaker. By these data we would conclude that if a man sounds gay, he probably is.

The qualifications here are interesting. About 25 percent of gay men were rated well within the typical range of straight speakers. Furthermore, there was more than twice as much variation (in statistical language, variance) among the gay speakers than among the straight speakers. Clearly, not all gay men speak in a recognizable pattern.

There is something to the stereotype that many gay men speak in a characteristic way. What is that way? Although some laypeople might have sufficiently skilled ears to discern the precise differences between gay and straight male speech, I do not. In the present stage of research, therefore, I am collaborating with linguists (more specifically, phoneticists), who make their living by listening for, identifying, and studying such differences. My collaborator has said this:

> Generally speaking, I'd say that the vowels appear to be shifted in a direction that would suggest a more fronted articulation. (For women, the vocal tract is shorter and differently proportioned, and the vowel shift may reflect modeling on the characteristic patterns of women's vowels.) In addition, we think we may be hearing more careful or precise articulation. We also think we may be hearing an articulation of /s/ that has the tongue positioned more towards the teeth, as opposed to the alveolar ridge (which is that hard ridge behind your upper teeth).

For the non-linguists among us, there are three main ideas here. First, gay men might pronounce vowels with their tongues more forward and to the top of the mouths than straight men do. Second, gay men might speak more precisely, articulating the sounds that straight men pronounce lazily. (Try reading this sentence both in your normal speaking voice and then in a carefully articulated voice. Does the latter sound more gay?) Third, gay men might produce sibilant "s" sounds, sounding somewhat hiss-like. If true, this observation may account for the idea that gay men lisp. All these hypotheses are testable, and soon we will know whether they are correct. Whether or not these particular ideas are correct, something makes a large subset of gay men readily identifiable by the way they speak just a few words.

How do gay men come to speak differently than straight men? Consistent with the central theme of this book, the first hypothesis that comes to my mind is that gay men are speaking in a feminine manner. That is, to the extent that gay men have recognizable speech patterns, those patterns may be somewhat like those of women or girls. Both straight and gay people tend to label the gay accent as a "feminine" speech pattern. However, it is unclear at this time what the gay accent is, much less whether it is feminine. If gay men wanted to speak like women, the most obvious way to do this would be to speak in a higher pitch, the way that many transsexuals and drag queens (in their female persona) do. But no one thinks they hear gay men speaking in a higher pitch. The idea that gay speech is feminine speech remains a hypothesis, for now.

If it is true, then there are at least two ways that gay men could come to speak in a feminine way. One is through (perhaps unconscious) imitation. This is particularly plausible if the gay accent is acquired during childhood, when feminine boys are most likely to have strong wishes to be the other sex. Perhaps they are attending to the ways that girls and women speak differently than boys and men, and imitating the former. The alternative hypothesis is that human males and females speak differently in part because their brains are innately different (due to early hormonal influences, for example), and that the brain centers affecting articulation are somewhat feminized in gay men.

The gay humorist David Sedaris wrote a story about being treated by a speech therapist for a year during grade school. He wrote that virtually all the boys seen by the therapist were sissies, like him. If this is even partly true, it suggests that some features of the gay accent begin during childhood. However, both Ken Zucker and Richard Green, who have worked extensively with feminine boys, have told me that young boys do not show it. They think it begins during late childhood or adolescence. This is an unusual time to acquire an accent, and it raises the possibility of cultural influence. If my 10-year-old son knows what it means to sound gay, so can these feminine boys who are becoming gay men. Could they be embracing and expressing their future identity? If so, the gay accent might not be feminine at all. Rather, it might be the product of the same kinds of semi-random factors that, for example, make Americans living below the Mason-Dixon Line speak in a "southern drawl." If this is true, there is still an important and puzzling distinction between the acquisition of a southern accent and a gay accent: Southerners grow up amidst people who speak with a southern accent; gay men do not grow up amidst people who speak with a gay accent.

****

I often don't have to hear a man talk or know what he does in order to have a strong suspicion he's gay. Sometimes it's enough just to see him move. If I see a man walking and displaying serious hip action, or keeping his elbow in while moving his forearm around; or if I see him standing with arms crossed and hands on shoulders; or if I see him sitting and waving his hands around a certain way when telling a story, my "gaydar" is likely to go off.

In 1999 psychologist Nalini Ambady of Harvard University published a study suggesting that homosexual people do in fact move differently than heterosexual people. In this study, she recruited homosexual and heterosexual men and women (targets) to be video-taped. Then she recruited a second set of homosexual and heterosexual people (judges) to view photographs, 1-second videotape segments, or 10-second videotape segments of the targets and try to estimate the targets' sexual orientation using this information. The three types of visual stimuli differ in the amount of "dynamic information"—the amount of information about movement. Photographs obviously contain little information about movement (although they might conceivably give some clues about posture), 1-second clips a bit more, and 10-second clips the most. She found that people's accuracy in estimating men's sexual orientation increased with the amount of dynamic information. Furthermore, even when she removed all static information—clothing, hairstyle, and so on—by using a computer to generate only the outline of the targets during their 10-second clips—people could judge men's sexual orientation better than by chance alone. However, Ambady didn't try to identify the specific components of targets' movement that judges used to make their decision. Indeed, she refrains from even speculating that the relevant information has anything to do with femininity.

Our lab took a slightly different approach. We used the same subjects who were the targets in our gay speech study, and we videotaped them walking down the hall, standing briefly, and then sitting briefly while conversing with one of us. Rather than recruiting new subjects to rate the targets, we found an existing rating scale that had been developed during the 1970s to study "sex role motor behavior." The table on the right shows some of the items that raters considered. Two students in my lab watched the videotapes and rated the subjects using the scale.

The results were quite similar to the results of the gay speech study. There was a large difference between gay and straight men. Again, there was much more variation among the gay than among the straight men. Only one straight man exhibited marked feminine movements. Him aside, 40 percent of the gay men were rated as more feminine than the most feminine heterosexual man.

Masculine and Feminine Traits

Walking

Masculine	Feminine
Long strides, free knee action	Short strides, controlled knee action
Minimum hip movements	Pronounced hip movements
Foot placement – straddling a line	Stepping on a line
Arm movements from shoulder	Arm movements from elbow
Firm wrist action	Limp wrist action
Arms hang loosely from shoulders	Upper arms held fairly close to body

Standing

Masculine	Feminine
Feet apart	Feet together
Arm movements from shoulder	Arm movements from elbow
Firm wrist action	Limp wrist action
Hand(s) in pocket	Hands on hips

Sitting

Masculine	Feminine
Buttocks away from chair back	Buttocks close to chair back
Leg not crossed or ankle on knee	Legs crossed, knee on knee
Precise hand motions	Graceful hand motions
Arm movements from shoulder	Arm movements from elbow
Firm wrist action	Limp wrist action

One important difference between our movement study and our speech study is that we were able to use the same movement scale to score both men and women. Not surprisingly, the scale yielded a huge sex difference. Gay men scored in the direction of heterosexual women, although they were much closer to heterosexual men. Although we don't know yet whether a gay accent is a feminine accent, we can conclude that gay men move in feminine ways. And this starts early, at least for some gay men.

Richard Green videotaped some of his feminine and masculine boys, and some girls, wearing clothes that concealed their sex (a bathing cap, for example). Masculine boys were clearly distinguishable from girls; feminine boys were not. Some of the feminine boys were as young as four years old.

In our study of sexual orientation and dance, we asked whether dancers could distinguish gay and straight male dancers by the way they dance, and most responded that they could. They elaborated that gay men were more feminine, and perhaps more dramatic, in their movements.

✶✶✶✶✶✶✶✶✶

The main characters of the movie, *The Birdcage* (originally a French Film, *La Cage aux Folles*) are a gay couple. One of them is a very masculine man, and the other is a flamboyant drag queen. In the movie, they clearly take separate roles as husband and wife, and this is a common stereotype about gay relationships. In this chapter I have been arguing for the accuracy of some stereotypes about gay men. What about this one?

In 1995 I became interested in using personal advertisements to study gay men's mating psychology. One can learn a lot about what people want in mates by studying these ads. They cost money, for one thing, and when people have to pay for each word, they try to make every word count. When describing whom they're looking for, people often have a mixture of idiosyncratic desires ("likes opera" or "enjoys camping"), but when the same preferences recur in ad after ad ("tall, dark, handsome, and rich" or "attractive, sexy, and fit"), you know these are commodities that most people want. For example, psychologists have analyzed personal ads to show that straight men are much more concerned than straight women about a potential mate's looks; straight women are more concerned about resources and the ability to acquire them: income, wealth, ambition, a good job, and intelligence. You can also tell a lot about the mating market by the way advertisers describe themselves. Advertisers want to entice readers to answer their ads, and are sometimes quite creative in their self-description. So the self-descriptive adjectives also tend to be those that are highly valued.

When my lab first started looking at gay personal advertisements, we were struck by a couple of differences from straight ones. First, gay men's ads were much more explicitly sexual than straight men's were— I will explain why I think this is so in the next chapter. The other difference was that gay men's ads used many more words related to gender conformity and nonconformity, such as masculine, feminine, butch, femme, straight-acting, straight-appearing, and flaming. This suggested that these traits were important to many gay men, but how so? If gay men tended to pair off as in *The Birdcage*, we would expect to see both advertisements in which the advertiser described himself as "masculine" (or "butch" or "straight-acting" or something similar) and requested a "feminine" (or "femme" or "flaming") partner; and advertisements with the reverse pattern ("Flamer looking for butch guy….."). We would expect to see similar numbers of both types. In order to check our expectation, we looked at more than 2,700 personal ads placed by gay men. For each ad, we looked for gender-related words and we kept count of how often the advertiser: (a) requested a masculine partner, (b) requested a feminine partner, (c) described himself as masculine, and (d) described himself as feminine. Forty one percent of all the ads had some gender-related word.

What we learned suggested that *The Birdcage* is indeed fiction. When advertisers requested either masculine or feminine characteristics in a partner, they requested masculine traits 96 percent of the time. Furthermore, when they described themselves as masculine or feminine, it was masculine 98 percent of the time. Both what gay men seek and how they represent themselves suggest that they are massively biased in favor of masculinity. Or is it a bias against femininity? In all 72 ads in which an advertiser was explicit about what kind of gender-related trait he did *not* want, it was a feminine trait; "no femmes" was the most common request.

These results raise at least a couple of questions. First, if gay men are almost all so masculine (as their self-descriptions imply), why do they bother requesting masculinity in partners? After all, most personal advertisers don't waste money asking for someone with four limbs, because even if they have this preference, they can reasonably assume that it applies to almost everyone. The answer is—and this will not surprise most people who have answered a personal ad—that people sometimes misrepresent themselves in a favorable way. How often do advertisers describe themselves as having "below average looks," even though half the world should? This consideration, as well as everything I've discussed in this chapter, should make one skeptical about accepting the masculine self-descriptions of gay male personal advertisers.

A second question is less easily dismissed. Perhaps gay men who place personal ads are not representative. Perhaps their unusual characteristics or preferences are what necessitate placing such ads in the first place. Maybe most gay men love feminine men, and because feminine gay men are plentiful, they don't need to advertise for them.

To answer this question we did a second study. We made up mock "gay dating brochures," each of which profiled two competitors. Each profile had both a picture and a self-description of an ostensibly gay man. Some of the pictures were of very attractive men, others of average-looking men, and the rest were of men we considered very unattractive. One of the descriptions was:

> Good-looking masculine gay man in early twenties seeks partner for relationship. I am in shape and enjoy rollerblading, jogging, and tennis. I live in the city and would like someone with whom I can share everything from an exciting evening in town at the clubs to a relaxing day at the museum. My hobbies include traveling, being outdoors, and listening to music.

The other descriptions were similar. The key word in the description above is "masculine." A third of the time, that word was included in the description; a third of the time, "feminine" was substituted for it; and a third of the time neither "masculine" nor "feminine" was included. Each brochure contained one description with either "masculine" or "feminine" and one description with neither term.

We went to a gay-oriented bookstore, a gay gym, and a gay pride rally, and we asked gay men to look at the brochures and choose which person they would prefer to date. Most of those polled chose the physically attractive men in the brochures—no surprise, gay men like good-looking guys. But the raters also strongly preferred the brochures with the "masculine" self-description. Substituting "feminine" for "masculine" had about the same effect as substituting an average-looking man's picture for a very attractive one.

The idea that gay men want masculine partners may be surprising to straight people, but it is less so to gay men. Jaye Davidson, the actor who played the homosexual transsexual in the movie *The Crying Game* explained: "To be homosexual is to like the ideal of sex. Homosexual men love very masculine men. And I'm not a very masculine person." The gay (and flaming) humorist Quentin Crisp speculated about gay men:

> To understand what kind of man they most admire it is only necessary to guess what they wish they themselves were—young, frail, beautiful, and refined. Hence their predilection is for huge, violent, coarse brutes.

Whether or not Crisp's explanation—gay men want masculine men to feel more feminine— is correct, he recognized the preference.

When gay men say "No femmes," what is it, exactly, that they are eschewing? Gay men tend to be feminine in several ways, including their interests, their voices, and their movements. (Although it is unclear that the gay accent is a feminine accent, even gay men discuss it as if it is.) Do gay men dislike hairdressers, men who speak with a gay accent, men with limp wrists, or all three?

One relevant but surprising finding from our study of gay interests, speech, and movement patterns is that a gay man who acts feminine in one respect doesn't necessarily display other feminine traits. For example, gay men who sound the gayest do not tend to be the ones with the most feminine movements or the most feminine occupations. If our results are correct, then knowing that a gay man is a hairdresser tells you nothing about how he sounds or moves.

When I ask my gay friends about what feminine traits they dislike, they usually begin by talking about the voice. An older acquaintance related how once in a gay bathhouse, he was on the verge of having sex with a very attractive and muscular stranger, when the stranger spoke. "When he opened his mouth, a purse fell out. I got limp." But when I went to a Halsted bar with my gay graduate student, he was able to determine which men he would likely reject merely by watching them move. We don't yet really know what gay men mean when they say they dislike femmes.

This leaves the question of why. When I talk about this with other psychologists, the most common suggestion is internalized femiphobia—femininity has been punished so often by the straight world that gay men, too, come to hate it. This makes sense to me, but it is not the only plausible hypothesis. Another is that behavioral masculinity characterizes the prototypic man. If one is attracted to men, then one will be attracted to those with masculine behavior. The second hypothesis is less malevolent but more pessimistic than the first. The second hypothesis implies that femiphobia is not due to societal intolerance but is intrinsic to male homosexuality and is not remediable even by reforming straight society to make it less homophobic. It suggests that across time and place, gay men will desire masculine men, and thus, acknowledging their own femininity makes them feel undesirable. We don't know yet how universal the gay male preference for masculinity is, although most of my foreign gay friends say that it is true in their locales as well.

Earlier in this chapter I suggested that having been mistreated as feminine boys is not the only reason gay men tend to react uncomfortably to the implication that they are, or used to be, feminine. The other reason, which I hope is now obvious, is that gay men themselves dislike femininity, or at least they find it sexually unattractive. To call a gay man "feminine" is not only to say that he is a target of many straight men's ill will, but also that he is less attractive than he would be otherwise. It is certainly an unfortunate state of affairs that gay men tend to be feminine, tend to be less attracted to femininity, but tend to be stuck with each other. There are similar ironies in straight relationships. The designer of the universe has a perverse sense of humor.

★★★★★★★★★

Until 1973 homosexuality was considered a "mental illness." During that year, the American Psychiatric Association eliminated it from its diagnostic manual (the *DSM*)—immediately curing perhaps 2 percent of the American population.

One of the influential events preceding the removal of homosexuality from the *DSM* was the publication in 1957 by Evelyn Hooker of a scientific paper in which she compared gay and straight men's responses on the Rorschach test. The Rorschach is the test that requires people to say what they see in inkblots. Hooker found that psychologists couldn't distinguish the answers of gay and straight men. In those days, most psychologists thought it was an excellent test of mental health. In recent times, the Rorschach has fallen into increasing disfavor, and some of us think it is little better than reading tea leaves. So the fact that psychologists couldn't tell gay men from straight men based on their Rorschach scores is not very meaningful.

During the past five years several studies have been conducted by giving large numbers of subjects standard psychiatric interviews, including questions about sexual behavior. In these studies, gay men have tended to have more of certain psychological problems than straight men. Specifically, they have been more likely to be depressed and anxious. One study found that gay men were more likely to have a history of suicidal thoughts. This was true even when the comparison group was the gay men's own identical twins. Gay men are also much more likely than straight men to have eating disorders like bu-limia and anorexia.

We don't really understand where these differences come from, but there are at least two interesting possibilities. One is that gay men have hard lives, particularly due to societal stigma, and this causes un-happiness and mental problems. This is the explanation that people tend to accept automatically, particularly pro-gay people. However, it might not be true. One of the studies that found higher rates of men-tal problems among gay men was conducted in the Netherlands. The Netherlands is probably the most tolerant country in the world to-ward homosexual people. But gay men there were still about three times as likely as straight men to have been depressed during the past year.

Another possibility is that gay men's pattern of susceptibility to certain (but not all) mental problems reflects their femininity. The problems that gay men are most susceptible to—eating disorders, de-pression, and anxiety disorders—are the same problems that women also suffer from disproportionately. We don't really understand why women are about twice as likely as men to become depressed during their lifetimes. It might be that their social world is more depression inducing. Perhaps women experience more stress. Alternatively, there may be innate sex differences that make women more susceptible. No one knows.

Part of the explanation for the sex difference in depression in-volves basic psychological traits. In particular, the personality trait, neu-roticism, is one of the strongest predictors of depression. Neurotic people are emotional worriers, and they react poorly to stress. Women are somewhat more neurotic than men, on average, although the dif-ference isn't large. Gay men are also more neurotic than straight men, although they are slightly less neurotic than women. Of course, we don't know where the sex difference, or the gay-straight difference, in neuroticism comes from. But genetic studies of twins suggest that it could be at least partly inborn. Even identical twins separated at birth are similar in their degree of neuroticism.

Learning why gay men are more easily depressed than straight men might tell us why women are also. This knowledge would also obviously take us closer to reducing depression among gay men. Therefore, we keep our minds open until science makes us certain we understand the sex and sexual orientation differences in depression. Regardless of what we learn, nothing I have written means that we should return homosexuality to the *DSM* and again consider it mental illness. Many characteristics are associated with increased risk for certain problems. Men commit much more violent crime than women, but women are more depressed than men. Young people are more likely to be delinquent, old people to be demented. Samoan adolescents commit suicide at much higher rates than Americans do. But we don't conclude from these facts that being male, female, young, old, or Samoan is a mental illness. Rather, the problems are being violent, depressed, delinquent, demented, or suicidal. Homosexuality, per se, is not a problem.

****

Gay men might be feminine in another, sexual, respect. One of the sex acts common between gay men is anal sex, and that act requires one man to be penetrated by another. This situation is analogous to that between heterosexual partners, in which a man is nearly always the penetrator, and the woman the receptive partner. When two gay men have anal sex, is the receptive one taking a feminine role?

Not necessarily. Men possess a prostate gland, an erogenous zone which is often pleasurably stimulated during receptive anal sex. This often feels good. *Bend Over Boyfriend*, an instructional sex videotape showing straight couples how women can (with appropriate equipment) anally penetrate their male partners, was a bestseller. Among gay men, enjoyment of anal penetration might sometimes reflect only a desire to experience a particular type of sexual pleasure, rather than a desire to enact a "feminine" role. Perhaps being penetrated anally is one of the perks of being a gay man.

But not all gay men prefer being penetrated. Some prefer being the penetrator. In common lingo, penetrators are "tops," and receptive partners "bottoms." Interest in this distinction increased during the AIDS epidemic, because scientists discovered that the receptive partner in anal sex was at greatly increased risk for HIV. HIV researcher James Weinrich showed that gay men who preferred the receptive role in anal intercourse were more likely than other gay men to remember being feminine boys. Similarly, our lab found that gay men who like to be penetrated describe themselves as more feminine than other gay men, although the association was not a strong one.

In our analysis of personal advertisements, we also looked at whether advertisers requested or described a preferred sexual role. A slight majority of advertisers described themselves as "bottoms" and sought "tops." In their sexual roles, gay men tend to be feminine. I have heard the complaint that at every gay bar there are "1,000 bottoms looking for a top" (though our results suggest that this is an exaggeration). One gay acquaintance related the following story, which is not unusual among gay men: "I met this cute guy at the bar. He seemed so butch and like such a stud, but when I got him home, the first thing he did was throw his legs in the air." (Crisp discusses a similar story in *The Naked Civil Servant*.) A common saying in the gay world is "butch in the streets, femme in the sheets." Gay men recognize that there is more than one way to be masculine, and that penetrative masculinity doesn't have all that much to do with the other ways.

The susceptibility of gay men to HIV infection might be due partly to their preference for receptive anal sex, which I have suggested might be a feminine preference. The other major factor that has made them vulnerable, however, is their masculinity, and it is to gay masculinity that I now turn.

CHAPTER 5

Gay Masculinity

I am at a dinner party in honor of an eminent professor, who happens to be gay, and I notice my friend Ben conferring quietly with his boyfriend, Charlie. Charlie leaves the room, and Ben tells me what they were talking about. Charlie had met a man at the party that he found attractive, and he was asking Ben whether it was okay if he went home with the man to have sex. Ben assured Charlie that this was fine, provided that he stayed within the bounds of their agreement: they had to use condoms, and the liaison must be exclusively sexual, and not romantic. Ben chuckled bemusedly thinking of Charlie's hesitancy to believe that Ben was really OK with this.

When Ben's panel of gay men speaks to my undergraduate class, someone invariably asks them about their feelings concerning monogamy. The panel's composition varies a bit from year to year, but typically, a majority of the men say that monogamy is unimportant to them. A couple of years ago, Rick was the sole monogamous holdout on the panel. I later learned that the next weekend, Rick and his boyfriend had opened their relationship to include sex with other men in the form of threesomes. The idea of consensually non-monogamous sexual relationships is, to say the least, challenging to the typical Northwestern University undergraduate.

The panel is also asked about the number of sex partners they have had, and their answers always elicit gasps. All the men have had hundreds of sex partners. Ben correctly reminds us that it depends on what we mean by sex—gay men don't have vaginal intercourse much, and most of the sex acts are oral. Still, even using the broader definition of sex, the typical heterosexual Northwestern student finds it amazing that anyone has had so many partners.

There is a big sex difference in the way my students react to these revelations. The women tend to be horrified, while the men are often envious. Although the straight men in my class have difficulty imagining the degree of cooperation necessary to have hundreds of sex partners, they have less difficulty understanding the desire to do so.

By heterosexual standards gay men are sexually promiscuous. Although this was well known to those who paid attention, it became clear to everyone during the AIDS epidemic. In a 1981 study by the Centers for Disease Control and Prevention, AIDS patients with an average age of 35 years reported an average of 60 sex partners per year, or approximately 1,000 lifetime partners.

Of course, the more sex partners a man has, the greater his risk of contracting AIDS, so gay men with AIDS would be expected to have more sex partners than other gay men. But surveys of gay men that ignore HIV status also find large partner numbers. In a survey of gay men from San Francisco conducted in the 1970s (that is, pre-AIDS), the typical gay man had had more than 500 sex partners, the large majority of whom were one-time flings. Many of these experiences were anonymous—some took place in gay bathhouses, for example.

Gay and straight men's personal advertisements differ in this respect too. Gay men's ads are often very sexually explicit. ("Wanted:

Masculine top for fuck buddy.") Although straight men often have explicit sexual fantasies they want potential partners to fulfill, it is doubtful that including them in personal ads would be very effective in recruiting women.

Gay men are less likely to enter meaningful sexual relationships. Surveys have found approximately 30 to 50 percent of gay men to be attached at any one time, compared to approximately 75 percent of lesbians and even higher percentages of heterosexuals.

Social conservatives have taken facts like these as evidence for the decadent and perverse nature of gay men. I think they're wrong. Gay men who are promiscuous are expressing an essentially masculine trait. They are doing what most heterosexual men would do if they could. They are in this way just like heterosexual men, except that they don't have women to constrain them.

****

Evolutionary psychology seeks to explain some psychological sex differences according to their evolutionary function, and interest in casual sex has always seemed to me its most convincing story. The currency of evolution—what determines how successful one has been, evolutionarily speaking—is the number of offspring one leaves. In the unconscious quest to maximize one's reproductive output, men are constrained most by the number of women they can have sex with. If a man had the universal cooperation of women, he could leave thousands of offspring in his lifetime. But women do not, in general, benefit *at all* from having more than one sex partner. A single man with average sexual ability can impregnate a woman often enough to guarantee that she will have as many children as she can have. Thus, we would expect evolution to have made men much more interested than women in sexual variety and casual sex. Casual sex with a variety of women is potentially a reproductive bonanza for a man, because it means that many more children carrying his genes will be out in the world. In contrast, unless there is something very special about a casual sex partner, women stand to gain little and lose much by having casual sex with him. That is, they risk losing the chance for commitment from and investment by other men.

I will mention several evolutionary arguments in this chapter, so let me anticipate an objection: Gay men are an evolutionary anomaly—they generally don't reproduce and pass on their genes—so evolutionary arguments don't apply to them. In fact, I agree that homosexuality remains an unexplained evolutionary paradox. However, many of the adaptations that evolved due to heterosexuality are also found in gay men. For example, gay men have penises, even though they don't use them for the purpose for which penises evolved, namely, procreation.

Regardless of whether one buys the evolutionary story, anyone who has been out of the house, much less to a singles bar, knows that men and women differ considerably in their approach to casual sex. Not all men want casual sex, and some women do, but the average difference is large. Determining just how large is difficult, because in measuring someone's casual sex interest our first impulse is simply to count his number of sex partners. The problem with this approach is that casual sex accomplishments are constrained by opportunity. In order for a heterosexual man to have casual sex, he has to have a willing female partner. If women aren't built to be eager to jump into bed with strangers, then straight men's numbers of sex partners underestimate their interests. On average, heterosexual men and women must have the same number of sex partners. This is not true of gay men, who do not need to negotiate with women in order to have sex. Gay men have sex with other men, who have similar inclinations. These inclinations include interest in sexual variety, and in having sex without commitment.

I devised a questionnaire that tapped interest in casual sex and sexual variety without asking for numbers of sex partners. The items in the questionnaire were written so that either straight or gay men or women could answer them.

INTEREST IN CASUAL SEX

Rate your agreement with each item, from 1 (strongly disagree) to 7 (strongly agree)

1. *I would consider having sex with a stranger, if I could be assured that it was safe and s/he was attractive to me.*

2. *I like the idea of participating in a sex orgy.*

3. *I would not enjoy sex without any emotional commitment at all. (This item is reverse scored.)*

4. *I do not need to respect or love someone in order to enjoy having sex with him/her.*

5. *I can't imagine spending the rest of my life with one sex partner.*

6. *Sometimes I'd rather have sex with someone I didn't care about.*

7. *Monogamy is not for me.*

8. *I believe in taking sexual opportunities when I find them, as long as no one gets hurt.*

9. *I could easily imagine myself enjoying one night of sex with someone I would never see again.*

10. *If an attractive person (of my preferred sex) approached me sexually, it would be hard to resist, no matter how well I knew him/her.*

I gave the questionnaire to both gay and straight men, as well as to lesbians and straight women. I also asked how many sex partners they'd had. Although gay men had had far more sex partners than straight men, they scored nearly identically on the questionnaire measuring interest. The number of partners of straight men was similar to those of straight women and lesbians. These facts suggest that women are responsible for the pace of sex. Gay and straight men both want casual sex, but only straight men have the brake of women's sexually cautious nature to slow them.

One e-mail listserv I subscribe to has a number of conservative subscribers. Recently, the topic of gay men's sex lives came up, and the

(mostly male) commentators balked at the idea that they would behave as gay men do if women let them. I think it is instructive to think
about how many opportunities the typical heterosexual man gets to
have uncommitted sex with attractive strangers. Probably not many
unless he is a rock star or movie star or looks like Brad Pitt. It is easy to
avoid opportunities that aren't there.

The evolutionary anthropologist, Donald Symons, who first used
the example of gay men to draw inferences about the male psyche
wrote, "I am suggesting that heterosexual men would be as likely as
homosexual men to have sex most often with strangers, to participate
in anonymous orgies in public baths, and to stop off in public restrooms
for five minutes of fellatio on the way home from work if women
were interested in these activities. But women are not interested."

★★★★★★★★★

Gay male couples tend to go through a typical sexual progression.
At first, when the relationship is new, sex is hot and heavy, and the
couple is happy being sexually exclusive with each other. But with
most couples, as the newness wanes so does the sexual excitement, and
eventually one or both partners seek sex elsewhere. In a study of 156
gay male couples in the pre-AIDS era, David McWhirter and Andrew
Mattison found that most became nonexclusive within a year, and all
were non-monogamous within five years. This pattern often occurs
even as partners become increasingly committed to each other in other
ways—emotionally and financially, for example.

The opening of gay relationships to outside sex partners undoubtedly reflects the compromises necessary for two men to fulfill
their masculine desire for sexual variety. But it may also represent a
feminine characteristic that I have neglected to mention. Many straight
men admit that they would love to have many more sex partners than
they do. However, few of these men would be eager to allow their
wives the same freedom.

Straight men tend to be sexually jealous. Gay men tend to be less so. One way to show this is to have people imagine their partner both having sex with someone else and falling in love with someone else, and then to choose which scenario is more disturbing. Straight men tend to find the sexual infidelity scenario more disturbing. Straight women and gay men tend to have the opposite reaction.

No one really knows why gay men react this way. It may simply be the compromise they have to make in order to have as many sex partners as they like. Perhaps both women and gay men recognize that their male sex partners are capable of having meaningless sex with others. Although this might not make either women or gay men happy, it is preferable to their partners having *meaningful* sex, because meaningful sex can more effectively end relationships. Or gay men's decreased sexual jealousy might reflect a more fundamental feminine trait. When straight men learn that their wives are unfaithful, they often become violent. Sexual jealousy is a major cause of domestic abuse and spousal homicide, committed most often by men. However, gay men are less physically aggressive compared with straight men. If sexual jealousy and physical violence are intertwined, this could explain why gay men are less sexually jealous.

★★★★★★★★★

In the course of my research on differences between gay and straight men, we also took photographs of our subjects' faces. In contrast to our results for speech and movement, raters could not tell gay and straight men's faces apart. This does not mean that gay and straight men look the same, however.

I work out at two gymnasiums, one on my university's campus, and the other in Boy's Town. The gay men at the latter are distinguishable from the mostly straight male students at the former. The gay men are noticeably more muscular and have less body fat. When they leave the gym, they are wearing tighter clothing, which reveals their honed bodies. Most of my female friends insist that gay men are better looking than straight men, and perhaps they are, overall (i.e., counting body as well as face). Why might this be?

If gay men are better looking than straight men, it is because they have to be. Gay men are masculine in the emphasis they place on a partner's physical attractiveness. Consider the questionnaire on the importance of a partner's physical attractiveness. When I gave that questionnaire to straight and gay men and women, the women scored lower than the men. Gay and straight men scored similarly. (Before we accuse men of being shallow, consider that straight women are more concerned than men about a potential partner's status and income.)

IMPORTANCE OF PARTNER'S PHYSICAL ATTRACTIVENESS

1. It is easy to imagine becoming romantically involved with someone I initially felt was physically unattractive, as I grew to know her/his personality. (reverse scored)

2. Looks aren't that important to me. (reverse scored)

3. In the past, I've usually initially become romantically interested in someone largely due to his/her physical characteristics.

4. It is more important to me how nice a potential romantic partner is than how good looking he/she is. (reverse scored)

5. I wouldn't consider being romantically involved with someone who was significantly overweight.

6. It would be hard for me to get involved with someone with a noticeable skin problem.

7. I like my romantic partner to dress attractively, even if it requires some effort on her/his part.

8. I would be upset if my partner did not try to maintain her/his physical appearance.

9. If my partner became much less physically attractive, it would be difficult for me to stay with her/him.

10. I would be happy if my partner were more sexually attractive than I.

Straight men have it easier, in this sense, because women are less hung up on looks and, therefore, easier to please. Gay men are victims of their own choosiness.

I have known few straight men who openly worried that they were not good looking enough for a successful dating career. In contrast, I know gay men who do worry, despite the fact that (in my opinion) they look fine.

It is relatively easy to make one's body attractive, and many gay men spend hours a week at the gym. Gay men are less likely than straight men to be obese. (Along with "no femmes," gay advertisements often add "no fatties.") But sometimes gay men go too far in the pursuit of thinness. Gay men are markedly over represented among men diagnosed with eating disorders, comprising 20 percent to 50 percent of male cases. (Remember, only 2-4 percent of men are gay.).

Gay men also tend much more than straight men to pay attention to fashion, such as designer clothes and trendy shoes. I don't think that this stems from the same roots as the desire to be muscular. I doubt that gay men care all that much whether a potential sex partner is stylishly dressed, as long as he is otherwise attractive. Rather, the interest in fashion appears to be a feminine trait. Gay men who are most interested in fashion tend to have other feminine interests as well.

****

The humorist, Cynthia Heimel, wrote that the most important male sex organ is the eye. This is equally true of gay and straight men, and in two senses. The first sense is the one just covered, the concern with physical appearance. The second sense is more overtly sexual. Both gay and straight men enjoy looking at naked people (of their preferred sex) a lot more than women do.

The questionnaire that asked about interest in visual sexual stimuli got results that differed markedly between men and women, with men scoring higher. This is one of the largest sex differences in human mating psychology (although it is much smaller than the answer to the question, "Do you prefer to have sex with a man or a woman?"). It also has a plausible evolutionary basis. A naked woman is potentially a sexual opportunity, and so straight men should have evolved to seek them out and to become sexually excited in their presence. In contrast, women have nothing to gain by becoming easily aroused at the sight of a naked man. Otherwise, men would be constantly exposing themselves to women and preventing women from using their good cautious judgment.

Although difficult to document, consumers of pornography are certainly disproportionately male. Even visual erotica that is targeted at straight women is disproportionately consumed by men—gay men. *Playgirl* magazine, which was intended to let women share in the objectification of the nude body, probably has as many gay men as women readers.

Gay men also respond to pornography very much as straight men do, and very differently from how women do. Show men two erotic video clips: one showing only men and the other showing only women. If they are straight, they become much more sexually aroused by the clip showing women than by the one showing men. This is true whether you measure sexual arousal by asking them (subjective arousal) or by using a penile plethysmograph, which measures the degree of penile erection (genital arousal). Gay men show exactly the opposite pattern: they are much more strongly aroused by video clips of men than by those showing women. Men are quite specific about the kind of erotic categories (i.e., male versus female) they respond to; we call this pattern "category specificity." In contrast, women of all sexual orientations tend to be aroused by video clips showing men and by those showing women. Women have a bisexual pattern of sexual arousal, both subjectively and genitally.

Men's category-specific pattern of sexual arousal is probably important in developing their sexual orientation. The experience of intense sexual arousal to one sex or the other, but not to both, is a powerful source of information.

INTEREST IN VISUAL SEXUAL STIMULI

1. *Seeing attractive people nude doesn't sexually arouse me. (reverse scored)*

2. *It would be exciting to watch 2 people have sex.*

3. *Seeing attractive people (of my preferred sex) in skimpy clothing such as lingerie or tight briefs is very sexually exciting to me.*

4. *I find photographs of attractive naked bodies (of my preferred sex) sexually exciting.*

5. *Being around a group of attractive naked people (of my preferred sex) does not sound very sexually arousing to me. (reverse scored)*

6. *When I meet someone I find attractive, I fantasize about what they would look like without clothes on.*

7. *Seeing the genitals of an attractive person (of my preferred sex) would be extremely sexually arousing.*

8. *Seeing my sexual partner undress is a real turn-on.*

9. *Whether or not I approve of them, I find films of attractive people having sex to be very sexually exciting.*

10. *When I see someone especially physically attractive, I may follow them briefly to get another look.*

11. *When I fantasize about having sex with someone, I try to picture very vividly in my mind what their body would look like.*

12. *If I had to choose, I'd rather have a long conversation with someone I'm attracted to than see them naked. (reverse scored)*

What about bisexual men? Although there are clearly men who call themselves "bisexual" and who have sex with both men and women, both scientists and laypeople have long been skeptical that men with bisexual arousal patterns exist. Kurt Freund, who invented penile plethysmography, related that he was never able to find a subset of men who appeared bisexual in the lab. Although their data are less scientific, gay men share Freund's skepticism. They have a saying:

"You're either gay, straight, or lying." In contrast, many women are bisexual; perhaps most are, at least in their sexual arousal patterns.

The motivation to seek erotic stimuli, such as strippers or pornography, and the tendency to be sexually aroused by erotica depicting one sex or the other (but not both) are characteristic of male sexuality. In these senses, gay men are masculine.

****

One of the first times that the gay panel spoke to my class, I offered to take them out for drinks on Halsted Street. I suggested that we go to Roscoe's, the only bar I had visited. Ben and his friends looked at each other knowingly and suggested that we go to Cocktails instead. Later they told me when they were about 30 years old, they felt too old to go to Roscoe's. I thought that they meant that the youngsters at Roscoe's were too unsophisticated to spend time with, but it became clear that they felt unattractive to the Roscoe's crowd. I am about 10 years older than they are so I wondered whether the boys (whom I had considered men when I was there before) at Roscoe's had been horrified by my presence there.

All the gay men with whom I have ever raised the issue believe that gay men are extremely youth-conscious. Even my 22-year-old research assistant has started to pine for his days in the sun when he was desirable to all gay men. The lonely aging queen, with no chance of a date, much less a relationship, is a common stereotype among gay men. Is this belief justified?

If it is justified, then it reflects another way that gay men are like straight men—in being most sexually attracted to younger adults. This contrasts with women, who tend to prefer older mates. This sex difference also has a fairly compelling evolutionary explanation. Men should have evolved to be attracted to women, especially when appraising potential long-term partners, who are fertile and will remain so for a long time. Because women's fertility is closely linked to youth, men should be more attracted to younger than to older women. Men's fertility does not decline much with age, and in fact, older men have acquired some of the things that women should value in a mate (including resources and evidence of survival skills). This could explain why women prefer somewhat older men.

I wrote the questionnaire on age preference to assess preference for younger versus older partners. In two separate studies gay men have scored intermediately between straight men and women. That is,

PREFERENCE FOR YOUNGER PARTNERS

1. Facial wrinkles in a potential romantic partner would be a real turn-off to me.

2. I find gray hair to be somewhat sexy in a potential romantic partner. (reverse scored)

3. I could imagine being romantically and sexually involved with someone 20 years older than I. (reverse scored)

4. If I had to choose someone other than my current romantic partner as a long-term romantic partner I would choose someone age _____. (reverse scored)

5. If I had to choose someone other than my current romantic partner as a one-time sexual partner I would choose someone age_____. (reverse scored)

6. I am turned off by bodies that show signs of aging (such as sagging skin or varicose veins).

7. I am most sexually attracted to younger adults (aged 18-25).

8. If someone showed definite physical signs of aging, it would be difficult for me to be very sexually attracted to them.

9. I find attractive adolescents (aged 16-18) particularly sexy.

10. I would be comfortable having a mate considerably older than I. (reverse scored)

11. It is hard for me to understand why anyone would have a strong preference for younger rather than older partners. (reverse scored)

they are less youth-obsessed than straight men but more so than straight women; they are a bit closer in this way to straight men than to straight women. Evolutionary psychologist Doug Kenrick has argued that gay men are identical to straight men in this regard. He has focused on the actual ages that gay men request in potential partners (in personals ads and in questions 4 and 5 on my questionnaire) and finds that these ages are quite similar to those preferred by straight men.

Whether or not gay men are just as youth-centric as straight men or are slightly less so, it is clear that aging gay men have a harder time finding desirable sex partners than straight men do. In this sense gay men have the same experience as straight women. In another way, however, their predicament is more difficult than that of straight women. Straight women actually prefer slightly older men, and certain traits such as accomplishments and wealth can make older men more attractive. (My sense is that, in contrast to women, gay men are not much more attracted to wealthy and accomplished men, but some are willing to trade company and sexual access for money and resources.) Straight women seem to value these non-physical traits more than gay men do. Most gay men would prefer partners in their 20s. Few women over 30 have the same preference. Therefore, it must be harder for older gay men to get what they want.

This does not mean that older gay men are doomed to be unhappy any more than it means that older women are. The idea that aging gay men are lonely and unhappy has been challenged by scientific research, which suggests that even unpartnered gay men tend to have busy and satisfying social lives. Evidently, older gay men do not focus as much as younger gay men do on sexual gratification. That is another sense in which they are like straight men.

★★★★★★★★

One main reason many parents express disappointment when they learn that their son or daughter is homosexual is that they fear they will not have grandchildren. However, there is a baby boom among homosexual people these days. Among lesbians, that is.

Obviously, it is easier for lesbians to become parents than it is for gay men. Lesbians can find sperm donors. At the very least, gay men must find women willing to donate their bodies for nine months to carrying a child. Still, I think that there is another reason why we haven't seen flocks of gay men seeking to become parents. (To be sure, many gay men became parents in the context of heterosexual marriages, though this seems to be getting less common.) It is because like straight men, gay men are less interested in children than women are.

When I teach undergraduate Introduction to Psychology, I sometimes show video clips of infants and children. I watch my students' faces, and the sex difference is striking. The women make googoo eyes and have smiles that betray profound desire. Although the men also smile, their smiles are unremarkable. For more objective evidence, consider the items in the questionnaire that I wrote to assess interest in children. In two studies, gay men scored similarly to straight men, but lower than straight women (whose scores were similar to those of baby-booming lesbians). The sex difference was not especially large, and this is one trait for which I suspect that subjects are not being completely honest in their responses. After all, who would deny experiencing warm, positive feelings from babies?

INTEREST IN CHILDREN

1. *I greatly enjoy spending time with young children.*
2. *I get a lot of pleasure from holding babies.*
3. *I would enjoy taking care of a baby for a friend or relative.*
4. *I daydream about having a baby of my own.*
5. *Often when I see babies, I experience warm, positive feelings.*
6. *When I think about it hard, I have strong doubts whether the rewards of raising an infant are worth the work and responsibility. (reverse scored)*

Although I am suggesting that neither gay nor straight men have a female-typical level of nurturance, I am not implying that men don't make good parents. Most men become devoted to their own children. And when men are faced with the challenge of raising children on their own, they usually rise to the task. But on average, men, compared with women, are less driven to have and raise children. So maybe we'll never have a gay male baby boom.

****

As the gay rights movement has challenged more and more traditional practices and assumptions, some gay writers have begun to challenge the prohibition of gay marriage, partly on the grounds that forbidding gay men to marry forces them into some of the behaviors that make straight people uncomfortable—most importantly, promiscuity. Writers like Andrew Sullivan and Bruce Bawer have argued that gay male promiscuity and relationship instability are primarily consequences of social stigma and society's unwillingness to formally recognize gay relationships. Undoubtedly, conferring such recognition would make gay male relationships somewhat more durable. Legal ties are inconvenient to break, and if gay couples could receive the perks given to heterosexual married couples (e.g., health coverage of spouses), that would also provide a positive incentive to stay together. However, discrimination cannot be the whole story. If it were, gay men and lesbians would have similar behavior patterns, but they do not. Among other differences, lesbians are more likely to be in relationships, place less emphasis on sex in their relationships, have many fewer sex partners, and are more monogamous.

Because of fundamental differences between men and women, the social organization of gay men's sexuality will always look quite different from that of heterosexual men's. Regardless of marital laws and policies, there will always be fewer gay men who are romantically attached. Gay men will always have many more sex partners than straight people do. Those who are attached will be less sexually monogamous. And although some gay male relationships will be for life, these will be many fewer than among heterosexual couples.

Social conservatives will view this prediction as tantamount to an admission of the inferiority of the gay male lifestyle, but it is not that. Nor do I wish to take the radical position that the traditional heterosexual family is sick, outmoded, or otherwise doomed. The aspects of gay men's relationships that cause discomfort—the preeminence of sexuality, the relatively short typical duration, the sexual infidelity—are indeed destructive in a heterosexual context, but they are much less so among gay men. There are two main reasons for this difference. First, gay male couples do not often have children, but heterosexual couples usually do. The main reason we strive for commitment in our sexual relationships is because we want to keep families with children from breaking up. This function is irrelevant to most gay men. It is relevant to those few who raise children, but they are unlikely ever to comprise a substantial proportion of gay men.

Second, men feel much less psychic conflict than women about casual sex. That is, most women are not only less interested in casual sex but actively avoid it, because it makes them feel cheap and used. In contrast, most men don't mind being used sexually, and whether or not there is emotional attachment, sex with an attractive partner is a self-esteem booster. When straight men try to engage women in relationships based solely on casual sex, they often resort to deception because they know that their prey want something else. Gay men are much less likely to practice this kind of deception. They are straightforward in their desires. This is the primary reason gay personal ads are so explicit. Does anyone doubt that straight men would specify their anatomical and sexual preferences if they could get away with it? The emotional consequences of promiscuity are much less damaging to gay men than to straight women. I am ignoring the obvious point that careless promiscuity can have awful health consequences for gay men. Given proper precautions, this issue essentially vanishes.

I suspect that regardless of the progress of gay rights, gay men will continue to pursue happiness in ways that differ markedly from the ways that most straight people do. This will be true even as society becomes increasingly tolerant of them. Both heterosexual and homo-sexual people will need to be open minded about social practices common to people of other orientations. They will also need to be cautious about recommending solutions to life problems that work for them but that might not translate well to the other camp.

CHAPTER 6

Danny's Uncle

Danny Ryan has a gay uncle, Mark, Leslie's brother. When Mark came out to Leslie at age 40, she felt that it was more than a coincidence that Danny and he are close relatives. When their mother revealed that Mark had been a feminine boy, Leslie was even more convinced that there was a real connection between her brother's and her son's behaviors.

She doubted it had to do with experience. Danny and Mark had not spent much time together. Nor had they had similar rearing experiences. The only similarity that immediately came to mind was that Mark and Danny had both been shamed by their fathers for behaving in feminine ways, though she somehow doubted that caused their femininity to begin with. Danny's behavior seemed to emerge despite his environment, not because of it.

Although she had never liked genetic explanations of behavior, which she associated with conservatism and racism, Leslie had to admit that it seemed possible. When she talked to Mark about it, he said he believed that homosexuality was inborn. So did most of the other gay men she knew. She had read about a study purporting to find a "gay gene." Perhaps in children the gay gene is a "sissy gene," and both Danny and Mark had it.

When Leslie Ryan came to me to talk about Danny, the possible genetic link was something I felt capable of discussing. I had spent the better part of my career studying it.

****

I became a sex researcher in 1986 when I was a graduate student in clinical psychology at the University of Texas. Although my advisor was fond of me, he was ready to see me finish graduate school. He thought that I was dawdling, and perhaps I was. Ignoring the low pay, graduate school is a fine existence for those who like to learn. But I was also running up against the dissertation block. I had primarily studied genetic aspects of schizophrenia and IQ, both of which have spawned enormous literatures. I was feeling that it would be difficult to make a valuable contribution in either area. The interesting ideas I came up with had either already been done or were too difficult to research.

I was taking a graduate course in sexuality and came across an interesting theory about male homosexuality. This theory was stimulated by findings from studies on rats in which pregnant females are stressed, for example, by being shocked and confined. The male offspring of these rats tend to show a feminine pattern of sexual behavior. Typically, male rats mount, and female rats lordose, or raise their rear end when a male tries to mount them. But the males who were prenatally stressed often failed to mount, and under some conditions they lordosed. Now, there is a big difference between what gay men do and what feminized male rats do. Gay men mount plenty. Still, it seemed plausible that in humans prenatal stress might cause male homosexuality. Research by a German scientist, Gunter Dörner, claimed to show just that. In his study, the mothers of gay men born during (or right after) World War II had had very stressful pregnancies, much more so than the mothers of heterosexual men born at the same time. For example, one mother had been forcibly raped, and several had lost their husbands in the war. One weakness of Dörner's study was that instead of interviewing mothers directly, he interviewed the gay and heterosexual men about their mothers' pregnancies. He says that the men were supposed to ask their mothers and then report back to him, but there is no assurance that they did, nor that the mothers were completely honest with their sons. Another potential limitation to his study is the fact that World War II Germany was a very stressful place. Could the kinds of events that happened to the mothers of gay men in his study be sufficiently common in modern-day America to be a common cause of male homosexuality?

I decided to do a methodologically stronger version of Dörner's study for my dissertation. I had never researched homosexuality before, and I had only a few openly gay friends, so I recruited gay and straight men from Austin using advertisements. (I offered them the staggering sum of $5 for participating.) I had to ask permission to send questionnaires to the men's mothers. However, I assured men that I wouldn't mention to mothers that the study had anything to do with homosexuality. Both the men and their mothers were terrific ally cooperative, and I eventually got a larger sample than Dörner did.

I also had fun meeting my subjects, many of whom were intensely curious about the science of homosexuality. At first I was a little uncomfortable being a heterosexual man studying gay men. Not because gay men made me uncomfortable, but because of inferences people often made about why I was doing the research. Some gay people were suspicious about my motives, and questioned what my agenda was. I quickly learned to emphasize my genuine pro-gay feelings, although it was difficult to make a strong case why my particular study would advance gay rights. And people tended to assume that if I was studying male homosexuality, I must be gay. At first, briefly, this bothered me, if for no other reason than the idea that people had false beliefs about something I considered to be an important aspect of my identity. I quickly learned that I would have to get over this, and I did. I now find it amusing and even flattering when people assume I'm gay—flattering when they say I have a good sense of style. I know most of the scientists who do substantial research on homosexuality, and there is certainly a correlation between this research preoccupation and sexual orientation. Perhaps half of us are gay, a much higher percentage than would be expected by chance.

Back to my dissertation. I found no support for Dörner's theory that men become gay due to prenatal stress. Mothers of gay and straight men reported similar memories of stress during their pregnancies. However, I had asked subjects about their relatives' orientations, and here I did find something. Gay men said that about 20 percent of their brothers were also gay, and straight men said that about 1 percent were. These numbers were similar to those that had recently been obtained by Richard Pillard, a Boston University psychiatrist. Because I had always been interested in genetics, I was excited by the prospect of studying the genetics of homosexuality. Just after I finished my Ph.D., I contacted Pillard, and we began a long collaboration on this topic.

★★★★★★★★

Just because homosexuality runs in families doesn't mean we can conclude that there are "gay genes." Catholicism runs in families, but there is no Catholic gene. Traits can run in families for either genetic or environmental reasons. To decide which is more important, we have to do more sophisticated kinds of studies, including studies of twins. Identical (also called "monozygotic," meaning "one egg") twins are genetically identical, and they provide the most direct test of the power of genes. If we could separate identical twins at birth and raise them separately, in different environments, we could see how similarly they developed. We know that separated identical twins nearly always have the same eye color, because eye color is virtually completely genetic. We know that separated twins can be somewhat different in their personality, although they are far more similar than unrelated people would be. What about sexual orientation?

Two pairs of separated male identical twins have been written about in which at least one of the twins was gay. Both pairs were separated early in life and reunited in adulthood. In one of these pairs, the other twin considered himself bisexual and had had sexual relationships with men, although he was currently involved with a woman. If gay men are correct in their skepticism that male bisexuality exists, this second twin is probably gay. The other pair was especially fascinating. The twins met, by accident, in a gay bar. They became lovers. (Sexual attraction between biological relatives who were separated early in life and later reunited is much more common than attraction between relatives who were reared together.) So both pairs of twins were similar, or as geneticists say "concordant" for homosexual orientation, which suggests that there are gay genes.

Although identical twins reared apart provide the ideal genetic experiment, they are quite rare, and *gay* identical twins reared apart are rarer still. Unfortunately, mothers of twins are insufficiently dedicated to scientific progress to routinely give up at least one twin for adoption. This means that we will probably never have a large enough sample of separated identical twins from which to draw firm conclusions. Luckily, we don't have to have separated twins to make scientific progress. Another kind of twin study, called the *classical* twin design, relies on the fact that there are two kinds of twins. Fraternal twins ("dizygotic"—"two eggs") are not genetically identical, but are only as similar genetically as ordinary brothers. The logic of the classical twin study is this: If gay genes exist, then identical twins should be more similar than fraternal twins in their sexual orientations. Gay men's identical twins should have a higher rate of homosexuality than their fraternal twins should.

In 1990, just after I arrived at Northwestern University, I advertised in gay publications across the country, from Boston to Los Angeles:

Are you a twin, or
Do you have an adoptive brother?

If you are a gay or bisexual man, and you have a male twin (either identical
or non-identical); or if you have an adoptive or genetically unrelated brother,
we would like to talk to you about participating in a national study of the
development of sexual orientation.

"Adoptive brothers" are genetically unrelated males raised together
as brothers. For example, if a family that already had a son had adopted
me, he and I would be adoptive brothers. Adoptive brothers are an
interesting comparison group, because they share no genes, only a
rearing environment.

Gay twins and adoptive brothers called our office, and we set up
interviews. Some we conducted on the phone, and some I conducted
in person on a whirlwind tour through several states. In Houston, I
met a parade of gay twins and adoptive brothers in my hotel room. In
San Antonio, it was a college dorm room, in St. Louis a professor's
office, in Kansas City and Dallas friends' apartments, and in Boston a
psychiatrist's office. It was a grueling month-long trip, but it was also a
fascinating one, because of my subjects, whom I remember vividly.
Usually, I met only one twin per pair, although there were exceptions,
such as the Kansas City identical pair who entertained me with stories
of their childhood aversion to sports and the Dallas pair who convincingly denied childhood interest in dolls but rather had played baseball.
There was the adopted St. Louisan who discovered that his biological
father was also gay. There was the haunted Houstonite so ashamed of
his homosexuality that he could tell no one, but who suspected that
his heterosexually-married identical twin was also gay. A couple of
heterosexually married subjects did reveal that they were "bisexual"
without their wives knowing. Some subjects were very sick from AIDS.
These subjects were especially happy to be telling their stories—they
shared my belief that the study was a worthy goal. Although I was
paying subjects $25 at the time, few were doing it for the money. Gay
twins wonder even more often than other people about nature versus
nurture.

I got quite a bit of information from these men, and from their brothers, but the main goal was to find out their twins' (and adoptive brothers') sexual orientations. If something genetic is going on, then identical twins should have the highest rate of also being gay, followed by fraternal twins, with adoptive brothers bringing up the rear. This is exactly what I found: 52 percent of identical twins, 24 percent of fraternal twins, and 11 percent of adoptive brothers were also gay. These results are consistent with a substantial, but incomplete, genetic effect. Gay genes apparently exist.

Could I say anything more about genes for homosexuality, other than that they exist? I had measured twins' femininity/masculinity retrospectively, as childhood memories. I wondered whether feminine gay men might be the products of "nature" and masculine gay men the results of "nurture." If this were true, and if "nature" is equated with "genetic," then feminine gay men would have more gay genes than masculine gay men. This should be reflected in the rates of homosexuality in their identical twins; identical twins of feminine men should have a higher rate. But I didn't find this.

I did find something interesting about twins' femininity, however. Recall that not all gay men remember being feminine boys. Some were conventionally masculine. Among the concordant identical twin pairs—those pairs in which both twins were gay—twins from the same pair had strikingly similar memories of their own childhood behavior. In some pairs, both twins recalled being extreme sissies. In others, both remember being masculine. Others were intermediate. This similarity was limited to the concordant pairs. In the discordant pairs—one gay and one straight twin—you can't use the memory of one to predict the other. The most feminine gay man might have a straight identical twin who plays football for a living. Usually, in fact, straight twins recalled being much more masculine boys compared with their gay identical twins.

I also looked at whether having a gay twin has any effect on one's own sexual orientation. For example, having an identical twin come out to you might make you question your own sexuality. I asked twins about this, and almost always, twins said that they knew about their own sexual feelings before they had any inkling about their brothers'.

The study, coauthored by Richard Pillard, was published in 1991 in *Archives of General Psychiatry*, and it received a fair amount of media attention. I got a flood of letters, including letters of support (primarily from gay people and their families) and hate-mail, primarily from religious people who hate homosexuality. When I sent news clippings to my subjects, I received letters back from several families or loved ones notifying me that the subject had died of AIDS. I found these men especially admirable, for maintaining their curiosity in the face of death.

****

If sexual orientation were completely genetic, then genetically identical twins should always have the same sexual orientation, but clearly they don't. Environment must also matter, but what particular environmental factors make a gay man? When most people hear "environment," they automatically think of the social environment, such as how children are raised or what kind of interpersonal experiences they have. But to a geneticist "environment" means anything not encoded in DNA, and this includes biological factors as well as social factors: diet, germs, and random biological events might also affect development. For example, when one identical twin is born with a major brain defect such as anencephaly (lack of a cerebral cortex, or outer brain) or microcephaly (a very small cortex), the other twin is usually normal. Only environment can cause identical twins to differ, but the environmental problem in this case is biological.

What kind of environmental factor can cause genetically identical twins reared in the same family from birth—often dressed alike and given the same toys—to differ in their sexual orientations? It is a fascinating question that we haven't begun to answer well. One hint comes from the childhood behavior findings. When identical twins differed in their sexual orientation, the gay one tended to recall being much more feminine than the straight one. This means that the environmental factors that cause the twin differences are there early on, by childhood. Based on other things we know, such as studies of children with cloacal exstrophy (like Jason/Amanda from Chapter 3), I suspect that these factors operate in the womb. But I don't know what they are.

So far, only one environmental factor has been related to male homosexuality: birth order. Psychologist Ray Blanchard has found that gay men tend to be later born sons in a series of brothers. In other words, gay men have greater numbers of older brothers than straight men do. Blanchard has obtained this same result in more than 10 studies, so we can have confidence in it. He theorizes that the "older brother effect" is a biological one, caused by the mother's immune system, which reacts increasingly to a succession of male fetuses. This immune response affects brain development and, in particular, the sexual differentiation of the brain. Blanchard is in the early stages of testing the immune hypothesis, so we don't yet know if it's true. And there are numerous firstborn gay men, so obviously birth order can't be the whole story.

Some social conservatives, especially fundamentalist Christians, have often worried that sexual orientation is affected by societal attitudes. Let people believe that it's okay to be gay, and we'll have more gay people. Western attitudes toward homosexuality have surely undergone a profound liberalization in the past 50 years. Gay people have become more visible, but there is no good reason to think that they're more common. Surveys of sexual behavior in the 1990s caused a stir because the prevalence of homosexuality was much lower than people had been assuming. (Before then, gay activists frequently asserted that ten percent of the population was gay. These surveys suggest that the true figure is more like one to three percent.) I suspect that there are now many fewer men who fight their homosexual feelings to live heterosexually than there used to be. I don't think this means that the trait of homosexuality has increased, although its expression may have.

Another environmental hypothesis that has received some enthusiasm on the political right is the idea that gay people seduce and recruit. I have never understood how this is plausible. A boy with no interest in homosexual activity wouldn't find it pleasant even if he agreed to it. How is this supposed to turn him gay? But recently, well-known scientists have presented findings they claimed support this possibility. Their study consisted of interviewing men about their earliest sexual experiences. They found that gay men were more likely than straight men to have had homosexual experiences in childhood and early adolescence. They concluded from this that the homosexual experiences actually caused the later homosexuality in these cases. I find this reasoning to be astounding. Boys in late childhood and early adolescence are often capable of sexual feelings. Some of those with homosexual feelings are surely motivated to act on them. And some get the opportunity to do so. Most gay men had their first homosexual feelings long before they ever acted on them.

To contemplate environmental influences on sexual orientation, consider this thought experiment. Suppose your task was to find a gay man, with the constraint that you can't ask anyone directly about his sexual preference. You can ask about their families or attitudes or childhoods, though. The question "Were your parents especially tolerant toward gay people?" would be an ineffective one. The question "Were you a very feminine boy?" would be much better. For men with an identical twin brother, the question "Is your brother gay?" would be intermediate in effectiveness.

****

In 1993 biologist Dean Hamer intrigued both scientists and the general public with his scientific report that male homosexuality may be caused by a gene on the X chromosome. Hamer's study consisted of two components. First, in interviews a sample of gay men reported more gay male cousins and uncles on their mothers' than on their fathers' sides. (Danny's uncle Mark is on his mother's side, and so is consistent with Hamer's findings.) This is the same pattern as X-linked traits like color blindness and hemophilia. The second finding was based on genetic linkage, which is rather complicated. Hamer studied pairs of gay brothers. Essentially, he reported that gay brothers shared a piece of the X chromosome, called Xq28, more often than they would be expected to if homosexuality had nothing to do with genetics. He concluded that Xq28 probably contains a gene that affects sexual orientation.

Hamer's results remain intriguing but doubtful. They have not yet been independently confirmed, and another scientist was unable to repeat the genetic linkage results. My own lab was unable to repeat his result that gay men have more gay uncles and cousins on their maternal side. Because the results are so important scientifically if they are true, it is important to do a definitive study. This requires money. However, even people who agree with the scientific value of this research are ambivalent about funding it if it might lead to socially undesired results. This is a general problem for research on the origins of sexual orientation. I have often been asked: "Do you want to find out what causes us so you can cure us?"

Hamer's findings provoked special concern. It seemed possible that we could soon discover a gene for homosexuality. If so, then we could develop a genetic test. Parents then might select against male fetuses that carried the gene. Hamer himself worried about this and urged that scientists and politicians band together to prevent it. Eerily, a play called *Twilight of the Golds* soon opened—it had been written before Hamer's study was published—whose plot revolved around a pregnant mother who discovers via genetic testing that she is carrying a gay child. Her brother is gay, but she ends her pregnancy anyway with tragic results.

I certainly have no motive to change gay people or prevent them from being born. I suspect that the world would be a happier place if we had twice as many gay people as we do. Still, there seemed to be something hysterical about the scenarios and arguments being thrown about. For example, some of the people raising the specter of "murdering gay babies" were the same people who insisted that abortion is no one's business but the woman's who is considering it. These people have never tolerated the use of words like "murder" and "babies" in the abortion debate before. ("Fetuses aren't babies. And abortion isn't murder.") Why now?

I discussed the controversy with Aaron Greenberg, a philosophically-minded attorney. Greenberg's analysis was provocative, surprising, and correct. First of all, it is useful to isolate the role of abortion in the controversy, because abortion itself is so contentious. The issue is not whether abortion is acceptable. Instead, the real question is whether parental selection in favor of heterosexuality is acceptable. To focus on this question, we have to assume that whatever means parents will use to do this are, in themselves, morally acceptable. So, if you have any problem at all with abortion, assume that pregnant women can guarantee a heterosexual child by, say, taking a pill, or avoiding certain foods, or even by reading their children certain bedtime stories.

What would make avoiding gay children wrong? One possible answer is that it would be wrong if it were done due to the parents' irrational dislike of homosexuality. And in fact, opponents of parental selection assume that parents selecting against homosexual children would necessarily have bad motives for doing so. But there are other reasons besides disliking homosexuality that parents might have for preferring heterosexual children: the desire to spare their children the difficulties of societal intolerance of homosexuality, the desire to maximize their chances of having grandchildren, or the desire to have children like themselves in an important area of life: heterosexual marriage and reproduction. No one thinks that Christian parents' desire to raise their child as a Christian is evidence of their hatred of Jews or anyone else. And even if parents prefer a straight child for a bad reason—because they hate gay people—this doesn't make the act of having a straight child wrong, at least in the absence of any harm caused by the act. A white racist may prefer vanilla ice cream to chocolate because vanilla is white, but this doesn't make ordering vanilla wrong, because ordering vanilla ice cream doesn't hurt anyone.

So the next question is whether selecting for heterosexual children would cause any harm. Certainly being straight rather than gay doesn't harm the child itself. Would there be some less direct harm caused by parental selection for heterosexuality that would make that selection wrong? One common argument is that allowing parents to choose heterosexual children would validate or encourage social intolerance of homosexuality. But letting Christian parents raise Christian children doesn't validate or encourage anti-Semitism. And both our ability to avoid children with birth defects and our tolerance of children actually born with defects seem to be at historical high points.

The belief that studying the causes of homosexuality will eventually harm gay people is a highly speculative one. There is no good reason to dislike or desire to harm gay people, and so it is difficult to argue that good scientific studies or rational, open discussion will have that end. In fact, I think that the more we know about homosexuality, the better attitudes toward gay people will become.

Every day gay people suffer real harm—indignities and abuse committed by those with irrational prejudice. We can do a lot more good by focusing on this problem, and trying to solve it, than by speculating about the harm that science might cause.

****

Every time I lecture about gay genes, the smart kid in the classroom asks the evolution question: How do gay genes persist if gay men reproduce less often than straight men? It is an obvious question, a fine question, an unanswered question. Homosexuality might be the most striking unresolved paradox of human evolution.

An adaptation is a solution to an evolutionary problem. Moving organisms do better if they see, so in response to this need, they evolve an eye. Of course, scientists and philosophers know that evolution is not intentional in the way I just implied, but is rather a process of random variation culled by differential success. Our ancestors were those who survived and reproduced best. The vast majority of our ancestors were heterosexual. Heterosexuality is a paradigmatic evolutionary adaptation. The desire to have sex with members of the opposite sex helps people have sex that might result in offspring. The number of healthy offspring one leaves is perhaps the best indicator of evolutionary success.

Homosexuality is evolutionarily maladaptive. I think this is an undeniable fact, although gay-positive people (and I am one) tend to cringe when they hear words like these. "Evolutionarily maladaptive" sounds like an insult, but it isn't one. Lots of traits and behaviors that are evolutionarily adaptive are less than admirable: jealousy, selfishness, dishonesty, infidelity, greed, and nepotism are all easy to explain evolutionarily. In contrast, extreme altruism is evolutionarily puzzling. However admirable they are, people who sacrifice their lives for the good of genetically unrelated others do not pass their genes to future generations. If people like Mother Teresa were much more common than they are, evolutionary theorists would be faced with the evolutionary paradox of saintly self-sacrifice.

Some people object that many gay men have sex with women, too. Many even have children. This objection is true but misses the point. Gay men have far less sex with women than straight men do. This means during most of evolutionary history, before birth control, they would have had fewer children. Today gay men have children at about 20 percent of the rate that straight men do. This is a massive disadvantage. Even if gay men had 98 percent as many children as straight men did, gay genes would be eliminated from the gene pool. Yet gay men persist, being at least 1 percent of the male population.

Speculating about the persistence of gay genes has become something of a parlor game among evolutionary scientists. They seem to have an amusing or intriguing idea, write a little paper, and then return to pursuits they consider more important. Unfortunately, the persistence of gay genes despite their evolutionary disadvantage is too serious an intellectual problem to be solved this way. The proposed solutions of evolutionary science's best and brightest have been rather lame.

The most common idea suggested by laypeople is that homosexuality is the human species' solution to overpopulation. Scientists are unanimous that this cannot be correct. Even if in the long run we would be better off as a species with more non-reproducing individuals, genes that make people non-reproductive cannot stay around long. They will rapidly be overtaken by the greater short-term success of reproducers. This idea is a non-starter.

There is what I call "the kind gay uncle hypothesis." This is the idea that instead of reproducing directly, gay men spread their genes by investing in their nieces and nephews. Perhaps gay men help out with childcare or give their relatives money. The recognition that we have a genetic interest in our extended kindred is called *inclusive fitness*. The idea that gay men reproduce indirectly, through relatives, has received favorable attention by the famous evolutionary scientist E. O. Wilson and others, but it is almost certainly incorrect. To begin with, the numbers are implausible. Children share half our genes, but nieces and nephews share only a quarter. For each foregone child, a gay man would have to have two extra nieces or nephews. He would have to be a super uncle. But when we surveyed gay men and their attitudes toward family members, there was no hint that they were any more interested than straight men in childcare or investing in their nieces and nephews. Of course, one can always say that it might have been different a long time ago, but this is the kind of objection that makes evolutionary theory immune from scientific testing. Furthermore, if the conditions that kept gay genes in the gene pool are no longer true, then those genes should vanish.

Another possibility might be called "the artistic brother hypothesis." Perhaps genes for homosexuality have other effects, and these are evolutionarily advantageous in heterosexual people. Suppose that like height, gayness depends on several to many different genes. Suppose that to be gay, a man needs to inherit five gay genes. If he inherits only four, he is heterosexual with some gay-typical traits. Perhaps he is artistic or has a keen sense of style. If so, then gay genes might persist, because most of the time genes that can cause homosexuality are in straight rather than gay people—and in straight people, they are evolutionarily helpful. If this hypothesis were true, then the heterosexual siblings of gay men would have more children than most people, because they would have more gay genes. There are two problems with this idea. First, how plausible is it that any trait that a sibling might inherit would make up for gay men's greatly diminished reproductive success? Do artists really have more than twice as many offspring as other people? Second, although male homosexuality is associated with other traits, such as femininity, the association is restricted to the gay men themselves. That is, the straight siblings of gay men don't appear to be more feminine, artistic, or style-conscious than other heterosexual people.

Most other ideas offered to explain how homosexuality has persisted have no more going for them than these do. Scientists have been trying to solve the intellectual problem on the cheap, a few hours at a time, rather than devoting the years it will eventually take to discover the secret. Until they invest the effort and resources that the problem deserves, homosexuality will remain an evolutionary mystery.

****

If gay genes exist, how do they work? No matter what causes homosexual orientation, these causes must work ultimately by impinging on the brain. Any psychological or behavioral trait must start in the brain. The brain is the prime behavioral and psychological organ. Other parts of the body might affect behavior, but they do it through the brain. For example, men make testosterone in their testes, but these affect sex drive—and everything else they influence—by acting on the brain. So if gay genes exist, they affect brain development. In fact, environmental influences also work through the brain. This would be true even if homosexuality were explicitly learned—which of course it isn't.

If there is a sexual orientation center of the brain, then we might expect it to work this way. If it is masculine, then attraction to women results. If it is feminine, then attraction to men results. Thus, both straight men and lesbians would be expected to have masculine sexual orientation areas, and gay men and straight women to have feminine areas. So one way to proceed would be to find areas of the brain that differ between straight men and women and then see if they differ between straight and gay men.

Neuroscientist Simon LeVay was thinking similar thoughts in 1989. Up to that point, LeVay had had a distinguished career studying how the brain allows us to see. But while caring for his longtime male partner, Richard, who eventually died of AIDS, LeVay became disenchanted with the vision work, which he felt was not sufficiently well connected to his personal concerns, such as his identity as a gay man. He had become less interested in the neuroscience of vision and more interested in the science of sexual orientation. He saw a paper by Laura Allen and colleagues entitled "Two sexually dimorphic cell groups in the human brain" and became excited. The paper claimed to find two cell groups (nuclei) that were larger in men than in women: INAH-1 and INAH-3. This is just the kind of finding that a neuroscientist interested in sexual orientation hopes for. The hypothalamus has been well established to be involved in sexual behavior in both humans and other animals. For example, manipulating hormones in rats to make them show either masculine or feminine sexual patterns seems to work primarily by hormonal effects on the hypothalamus, in particular an area of the brain called the Sexually Dimorphic Nucleus (SDN). The SDN is much larger in normal male rats than in normal females. Males deprived of testosterone during a critical period of development fail to mount like normal males during adulthood, and they have small SDNs. LeVay thought that perhaps either INAH-1 or INAH-3 might play the same role in humans that the SDN does in rats.

In 1990 LeVay began his own study. He was able to obtain the brains of several gay men, all of whom had died of AIDS; the brains of several straight men, most but not all of whom had died of AIDS (these were IV drug users); and those of several straight women. In contrast to Allen's original study, LeVay found no difference between straight men and women (or gay men) in the size of INAH-1. However, in INAH-3 he struck gold. Like Allen, he found that INAH-3 was larger in straight men than in straight women. The finding that put him on Oprah, though, was that gay men's INAH-3s looked like those of straight women rather than straight men's.

This study, published in *Science* in 1991, was the first of the "Causes of Sexual Orientation" studies that made big splashes in the early 1990s, and like the others (including Hamer's and my own), LeVay's received a great deal of scrutiny. One big concern was whether AIDS had caused the difference—remember that all his gay men had acquired AIDS, but some of his heterosexual subjects had not. But LeVay showed that this was not the explanation. The size of INAH-3 did not differ between the brains of straight men who died of AIDS and those who did not have the disease. (It is aggravating to see this criticism continue to pop up in critiques of "gay science.")

A second objection that was often raised was that being gay might have somehow caused the difference in INAH-3, and not vice-versa. If this were true, then LeVay's finding would have no relevance for the cause of sexual orientation; it would represent an effect of sexual orientation. The problem with this idea is that the hypothalamus appears to develop early. Not a single expert I have ever asked about LeVay's study thought it was plausible that sexual behavior caused the INAH-3 differences.

The main thing that LeVay's study has had against it is that it has been difficult to repeat. Repeating interesting and surprising studies is important, because too often they turn out to have been false leads. In 1991 LeVay was in a good position to try to repeat his own study, but he elected to stop being a laboratory scientist and to focus on writing science books for the general public. Also, as Dean Hamer's example shows, scientists are less impressed when a scientist repeats his own results than when an independent research group does it.

LeVay's study soon got much more difficult to repeat. It was possible only because of the terrible casualties of the AIDS epidemic, which thankfully, have recently slowed considerably. Still, one dogged scientist has managed a reasonable attempt to both repeat the INAH-3 work and to take it further than LeVay did. I first met Bill Byne in 1991, just after Richard Pillard and I published our male-twin study. Byne was extremely skeptical of LeVay's results—in fact he believed they were probably incorrect. At the time, I thought that Byne's skepticism bordered on paranoia, because the concerns he kept raising seemed far-fetched. For example, he thought that perhaps AIDS caused by homosexual behavior might have different effects on the brain compared with AIDS caused by IV drug use. This seemed like a stretch to me—how does HIV know how it gets into someone's body?

Although I was wary of Byne, I also recognized his intelligence and appreciated his tenacity. Byne's work and ability have paid off recently in a series of important scientific findings. First, he repeated the finding, yet again, that INAH-3 is larger in straight men than in straight women. Even Byne thinks that the INAH-3 sex difference is no longer open to question. Second, he examined the biochemical composition of INAH-3 and showed that it was similar to the SDN in rats. This suggests that those structures may have similar functions, or at least similar evolutionary origins. Third, he looked at the brain of a female rhesus monkey that had been given prenatal androgens and found that this monkey had a larger INAH-3 than other female rhesus monkeys. (Byne did not kill monkeys for his study; their brains were being stored in a laboratory.) Finally, Byne has managed to assemble a sample of brains from straight women and gay and straight men, and he has repeated LeVay's study. His results were similar to those of LeVay, although the difference between straight and gay men was not quite as large. If I were going to invest a million dollars to look for the causes of male sexual orientation, I might well invest it in Byne and INAH-3.

****

It sometimes seems that there is a news story every other week linking some kind of biological trait to sexual orientation. I have seen reports that gay people have different fingerprints, more left-handedness, different inner ear waves, height, and even penis size. (Gay men allegedly have larger penises.) These findings are related to two different trends, one of which is good, and one of which is not. The good trend is that people are interested in sexual orientation, especially its causes, so whenever scientists report a finding, the press is eager to report it. The bad trend is that scientists and laypeople both think that there is something more impressive about biological markers—a trait that one can measure with a ruler or a blood test or complicated electronic equipment (I'm serious, there is no better definition of "biological marker")—than there is about psychological or behavioral markers. This is a mistake.

Let's take the recent example of 2D:4D. In women, the index finger (2D, second digit) tends to be similar in length to the fourth digit (4D). In men, the index finger is more often shorter than the fourth. The ratio of 2D to 4D, 2D:4D, is larger in women than in men, on average. Relatively speaking, this is a small difference—the effect size, d, is only about .30, which is the smallest effect size mentioned in this book. Scientists who study 2D:4D argue that the sex difference depends on prenatal androgens, such as testosterone. My first reaction is that if 2D:4D is so androgen-dependent, then why is the difference between men and women so small? But let's assume that prenatal androgens are at least one cause of 2D:4D. Sexual orientation would be a sensible trait to examine.

In 2000 a report appeared in the journal *Nature* that lesbians had masculine 2D:4D ratios. There was no clear finding for gay men. The chief author of the paper, Marc Breedlove, is a well-known and capable scientist. The report made all the papers and Jay Leno's monologue. As to be expected, given the modest sex difference, differences between lesbians and straight women were small, detectable primarily because the authors had collected a very large sample at a gay pride parade.

Recently, Richard Lippa collected finger length data from even more subjects at gay pride parades, and his findings disagree with Breedlove's. Lippa didn't find any differences between straight women and lesbians, but he did find a difference between gay and straight men. As we would expect, gay men had a feminine pattern (larger 2D:4D). Because Lippa's sample was much larger, and because his analysis was somewhat more sophisticated (for example, he analyzed results separately by ethnic group, because there are ethnic differences in 2D:4D), I think it is somewhat more likely that his results will stand the test of time. Conflicting findings are common in science, and eventually we will know whose findings are correct, if either is. But in any case, I don't think that studies of 2D:4D will reveal nearly as much about the causes of homosexual orientation as looking carefully at boys like Danny Ryan will. And let me make clear that I think we should be looking at Danny's behavior, and not his fingers.

Nobody ever gave Danny Ryan a dress before he asked for one, and he was punished much more than rewarded for his gender non-conformity. If he grows up to be a gay man, as I expect he will, it will be *despite* the most obvious influences in his social environment, not *because* of them. In the short term, Danny will receive no more encouragement from others to become gay than he did to wear dresses. Behavior that emerges with no encouragement, and despite opposition, is the *sine qua non* of innateness. Boys like Danny are poster children for biological influences on gender and sexuality, and this is true whether or not we measure a single biological marker.

CHAPTER 7

Is Homosexuality a Recent Invention?

What about the Greeks? This is the other question (with the evolutionary one) that is asked nearly every time I talk about homosexuality. It would be asked even more often, except that a lot of people who believe that they already know the answer to the question don't show up to my lecture. These people think that I am wasting my time trying to learn about the nature of homosexuality. They think that homosexuality has no nature. To them, homosexuality is a "social construction."

"Social constructionism" (or "social constructivism") is a term that might be familiar to anyone who has taken a humanities course at an American or European university since 1990, but it might otherwise sound odd. It is difficult to explain social constructionism in a way that satisfies social constructionists. They think this is because they are profound and people like me simplistic. I think it is because they aren't very clear, and to the extent they are clear, they are incorrect.

Regardless, this is an important controversy, and I will try to do it justice.

Across history and culture the social constructionists believe they see tremendous variation in both the prevalence and form of homosexual behavior. So much so that to talk about sexual orientation as if it might be a fundamental part of human nature is surely mistaken. They think that whether one prefers sex with men or women is more like whether one likes or dislikes opera than whether one has a penis or a vagina. Searching for scientific laws that explain who likes and dislikes opera, without paying much attention to the culture that one is in, is surely an absurd pursuit. Social constructionists think that those of us who study the origins of sexual orientation are, likewise, spinning our wheels.

Some of the cultural and historical phenomena that social constructionists have focused on include:

• Ancient Greece, where it was common for men to form sexual relationships with adolescent boys and where most men were bisexual (according to the constructionists).

• The Romans, who were tolerant of male homosexual behavior, provided that normal free men were penetrating male slaves and prostitutes.

• Fifteenth-century Florence, where nearly half of all men came to the attention of the authorities for committing sodomy.

• The Sambia, a tribe in New Guinea, in which boys live for years only with males and practice oral sex with men.

• British public schools, all-male boarding schools, which were famous for their high levels of homosexual activity.

What do these examples have in common? First, these cultures had much higher rates of homosexual behavior between men than exist now in the West. Second, the constructionists assert that none of these cultures thought about sexual orientation the way we do. Indeed, the social constructionists claim that the idea that people vary in whether they prefer to have sex with men or women began recently, during the nineteenth century. Before then, there were only homosexual and heterosexual acts, not homosexual and heterosexual people. Perhaps the most surprising claim the social constructionists make is that the way that cultures think about sex actually influences the sexual feelings of their members. In cultures that don't classify heterosexuals and homosexuals, men are capable of sexual attraction to both men and women.

Social constructionists call those who disagree with their major points "essentialists." Essentialists believe that sexual orientation is an essential part of human nature. I am an essentialist.

★★★★★★★★★

How can we know anything about the sex lives of Greeks who lived 2,500 years ago? It is difficult enough to know about the sex lives of Americans today, even though we can discuss the question with at least some current inhabitants of our culture. Furthermore, social scientists have completed several large and ambitious surveys on this topic during the past decade. These studies required millions of dollars and thousands of subjects. Typically, the researchers identified a representative sample (of Americans, for example), and telephoned them, asking them to participate in a sexuality survey. A standard interview was used that included questions such as: "How many same-sex partners have you had during your lifetime?" "Do you consider yourself heterosexual, bisexual, or homosexual?" "Are you attracted more to members of your own sex or to members of the opposite sex?" Interviews were conducted in private, and subjects were assured that their answers would remain secret. We know exactly what percentage of people declined to participate, and some studies even tried to get at the reasons for non-participation. Still, these studies did not eliminate all uncertainty about American sex lives. There remains controversy about how frequently people commit adultery, for example, and the estimated prevalence of homosexuality varies widely, from less than

1 percent to more than 4 percent. This is a wide range, with the higher estimate four times the lower one. The fact that we can't get complete accuracy is not all that surprising, given the sensitivity of the topic.

Historians have a much more difficult task. They can't go back in time to ask people what they feel and think. They have no access to surveys. Rather, they get remnants and pieces, from art, law, myth, fiction, graffiti, history, philosophy, politics, and poetry. They can make educated guesses. To the extent that different sources of information all convey a similar picture, we can be more confident. Still, there is no reasonable dispute that we can be much less certain of facts about Ancient Greek sex lives than we can about the facts of American sex lives in our own time. Historians' work is important, but it just can't be nearly as conclusive as even contemporary social science can be. The best historians are appropriately careful about the conclusions they draw. As one prominent historian put it: "A scene on a vase may not tell us any more about a middling Athenian than a Wedgwood china pattern tells us about a Victorian hackney driver."

Here is what we know for sure. In parts of ancient Greece, including Athens during the fifth century, B.C.E., some men formed homosexual relationships of a different type than that which commonly exists now in the West. Typically, these relationships involved an established older man and a younger adolescent boy, the younger partner (the "eromenos") being near the age of first growing a beard, the older ("erastes") often an older (bearded or lightly bearded) adolescent or unmarried young man. The older partner could also be a married man—marriage was largely non-companionate in Greece. The relationship was cemented by the older partner giving the younger a gift, often shown on vases as a cockerel (a young rooster). These relationships were sexual ones. The main sexual activity depicted on vases was intercrural intercourse, in which the older partner inserted his penis between the thighs (but not in the anus) of the younger partner, thrusting until ejaculation. Nobody on either side of the social constructionist debate denies that this practice existed.

But the most important social constructionists' claims have not withstood close scrutiny of historians and classical scholars. Not surprisingly, careful scholars admit that there is much we don't know about the sexual practices and desires of the Ancient Greeks. One problem is that Ancient Greece included many diverse cultures—Sparta, Athens, Crete, among others—and several centuries. To make any generalizations about "the Greeks" is risky. What we do know contradicts the social constructionist account. For example, far from being widespread, "pederastic" relationships between men and adolescent boys were viewed as a decadent practice of the aristocracy. Parents often tried to prevent their sons from entering these relationships (as the younger member). If money changed hands, the younger member could lose citizenship. The Greeks were especially intolerant of receptive anal intercourse, which they viewed as an abomination against nature.

More important, the record we have shows that some Greeks recognized that at least some people had a homosexual preference. For example, Aristophanes portrayed Agathon as a feminine man who enjoyed receptive anal sex. In Plato's *Symposium*, Aristophanes related a creation myth in which originally there were three sexes: men, women, and a combination of the two. Zeus cut each sex in half, and from that point, each person was driven to find the missing half. Thus, the man created by cleaving a complete man in half was homosexual, whereas heterosexual men and women were created by cutting the original androgyne in half. The historian John Boswell documented the existence of obviously heterosexual or homosexual characters in Greek literature.

The Romans, just a few centuries later, had a word to describe feminine, exclusively homosexual men: cinaedi. These men were so common that the Apostle Paul offered homosexual behavior as his chief example of the capital's decadence. They appear to have shared a flamboyant style of distinctive dress, hairstyles, and mannerisms, as well as regular cruising grounds, and typical occupations. To me, they sound a lot like the guys on Halsted Street.

****

Fifteenth-century Florence had a reputation as a bastion of "sodomites." (This is why "Florenzer" in German meant "sodomite.") In 1432 the city created a commission, "the Office of the Night," to solicit and investigate charges of sodomy. For example, boxes were placed so that people could make anonymous accusations. The population of Florence was 40,000, and the Office of the Night lasted 70 years. During that time, 17,000 men were implicated. Assuming there were 20,000 men in Florence at any one time, and that 70 years means two complete generations, 17,000 is nearly half of the men of Florence during that time. Florentines generally accepted sodomy as a common misdemeanor, to be punished with a fine, rather than as a serious crime. Fewer than 3,000 of the 17,000 accused men were convicted.

The primary historian of homosexuality in fifteenth-century Florence, Michael Rocke, emphasized the social constructionist line, that these men were not considered "homosexual." However, at least some clearly were. One man confessed to his acquaintance, Machiavelli, that had his father "known my natural inclinations and ways, [he] would never have tied me to a wife." There was a core minority of "notorious sodomites" who committed a disproportionate number of offenses. And accused men were more likely to be bachelors than married men. All this argues that some men preferred men to women.

Florence seems to be a special case at that historical period. Other similar-sized cities in the region don't seem to have had its high rate of homosexual behavior. There is some evidence that some men migrated to Florence precisely because of its reputation, making it a kind of Renaissance-era San Francisco. Thus, it is unclear to what extent the beliefs and culture of Florence influenced the sexual desires of its inhabitants. Perhaps the influence was primarily the other way around.

****

The Sambia live in an isolated forested area on the edge of the New Guinea Highlands. Sometime between ages 7 and 10, Sambian boys are taken from their mothers and made to live in an all-male house in their village. They will live there for the next 10 years, learning Sambian ideology and Sambian practice.

The ideology stresses the importance of semen, which is considered necessary for male virility and health. In contrast, women's essential substance is menstrual blood, which is potentially poisonous to men. Women are born with the ability to secrete menstrual blood, although they don't do so until adolescence. In contrast, males are born without semen, which they will need in order to maintain health, risk sex with women, impregnate women, and even help provide women with milk.

But how do males get their semen, if they have none at birth? They get their semen by sucking it out of older males. Older boys teach them how to do this. Although the older boys obviously get enough sexual pleasure to enable them to ejaculate, everyone considers them to be doing a favor by donating their semen to the younger boys. The younger boys who fellate the older boys don't usually say that they're sexually aroused. The younger boys suck semen from older boys a few times a week for several years (remember, they are still segregated from their families and all women) until they undergo puberty. This event "shows" them the value of eating semen. Soon, the mature boys will begin supervising the initiation ritual of a new crop of 7-10-year-old boys, now as the older, fellated, partner.

In a few more years, in their late teens or early twenties, these now young men are ready to be married to women. At the first stage of marriage, they don't have vaginal intercourse, but instead, have oral sex with their wives. They also continue to donate semen to younger boys. Soon, however, the men stop having homosexual encounters with younger boys and begin having vaginal intercourse with their wives. Most men make this transition, from completely homosexual behavior during adolescence to completely heterosexual behavior during adulthood, with no problem. A small minority of men remain bachelors and seems to prefer the boys. The other Sambia ridicule these men and think they are odd. We might call them "gay."

****

At a scientific conference I was talking to a British colleague about homosexuality. My impression of him was of a strict heterosexual, but he revealed that he had had numerous homosexual encounters as an adolescent at public boarding school. (In Britain, "public school" refers to the fancy kind of school that Americans call "private school.") This activity consisted of mutual masturbation, and he found it exciting at the time. He puzzled over his transition from those days to now, in which he found the idea of sexual contact with other men "disgusting."

His experience reveals three facts that I think might be important. First, his homosexual experiences occurred in a context of female unavailability. The same was also obviously true of the Sambia, but it was also true to a somewhat lesser extent of the Ancient Greeks, who sequestered women until marriage. Men seem more likely to resort to homosexual behavior if they have no heterosexual outlets.

Second, his enjoyment of homosexual behavior was evidently normal for him; his later disapproval required him to learn that homosexuality is disgusting. (He does not, however, believe he could have enjoyed fellatio or anal sex even at that time.) Perhaps, then, many if not most men could enjoy homosexual encounters of some form if they hadn't learned to be disgusted by them.

Third, there might be something to the idea that young or adolescent boys are especially attractive as homosexual partners for males. They are relatively hairless and less muscular; in these senses they look like women.

None of these possibilities destroys the idea that sexual orientation is a meaningful concept. To say that many men who don't have access to women will have sex with men is not to say that this is their first choice. That a high percentage of men might enjoy homosexual sex if they were not socialized otherwise does not mean that they wouldn't prefer heterosexual encounters given the choice. To say that men find relatively immature males more attractive than mature, bulky, hairy men is not to say that they prefer young males to females.

Sexual encounters in prison are illuminating. Men in prison generally have no heterosexual outlet, except through fantasy. It is remarkable that a high percentage of men who have always been heterosexual outside of prison engage in homosexual sex in prison instead of confining sexual activity to masturbation. Evidently, there is something pleasurable about being with another person that is more rewarding to some men than merely rubbing their own penises. Many, if not most, incarcerated men have the capacity to enjoy homosexual behavior. The most favored partners are the young, weak, and feminine. Although sociologists have tended to interpret this in terms of power relations—stronger inmates want to dominate weaker ones—it seems more likely to me that this is because the young, weak, and feminine make better female substitutes. There is one type of prisoner that prefers the big, masculine men as partners. This is the "queenie" prisoner, the feminine gay man, who receives a good deal of sexual attention from other prisoners.

When they leave prison, men who had been heterosexual before entering usually return to a strictly heterosexual lifestyle. Their prison encounters did not indicate that their sexual preference had changed. The men were simply doing the best they could, given constraints. It would be important to know what these men were thinking when they were having their penises sucked by other men, for example. Were they thinking of the men sucking their penises, or were they imagining their girlfriends at home? The former possibility would indicate more flexibility of true sexual preference than the latter.

The social constructionists have offered Greek pederasty, the Sambia, and the British boarding school experience as proof that men are innately bisexual. Although there may be something to their argument, it can't get them as far as they'd like to go. They can't explain the contemporary Western gay man. All these men grew up in a culture that gave homosexuality an inferior status compared with heterosexuality. Many of them were at some point highly motivated to be straight, and some tried—through prayer, therapy, or marriage. If bisexuality—meaning indifference to the choice between male and female sex partners—were in them, then it should have been easy for them to conform to the heterosexual norm. But it wasn't.

Recall gay men's skepticism about men who claim to be bisexual. ("You're either gay, straight, or lying.") My lab has been trying to find bisexual men by studying men's erections to male versus female sexual stimuli. A truly bisexual man should become substantially aroused to both sexes. Out of approximately 30 men who claim to be bisexual, only 2 have sexual arousal patterns that might be classified as bisexual. Most of the rest had a gay arousal pattern; a few had a straight pattern.

In the right culture, most men might be capable of some sexual arousal to both sexes. However, this doesn't mean that they wouldn't have stronger feelings for one sex or the other. Most would probably have much stronger feelings for women, and a minority would have much stronger feelings for men.

✶✶✶✶✶✶✶✶✶

Social constructionists have made a positive contribution to understanding sexual orientation by their insistence that we attend to different cultures rather than assuming everyone is like us. There are interesting differences between the types of homosexual behaviors and relationships across cultures, and we don't yet understand them fully. However, the contention that homosexual orientation (as distinct from homosexual behavior) is a recent and local phenomenon is not supported by the evidence. Men who look awfully similar to the men I've been talking about in previous chapters seem to have existed through the ages and in vastly different cultures. Social constructionists' refusal or inability to see this suggests that they are trying to keep their eyes closed.

****

Transgender homosexuality occurs when one man takes on a feminine role, often dressing as a woman and taking a woman's name, and has sex with masculine men. Transgender homosexuality is probably the most common form of homosexuality found across cultures. It occurs in the West (although it is much less common than egalitarian homosexual relationships), and has been documented in a number of other societies. One of the most extensively studied transgender homosexuality traditions has been that of the *berdache*, among some Native American tribes. The *berdache* tradition involved males who typically were identified in childhood by their femininity and placed in a role that would allow them both spiritual leadership and sex and marriage with men. As one observer said about the *berdache* among the Crow of the Plains in 1903: "I was told that when very young, those persons manifested a decided preference for things pertaining to female duties."

The *hijras* of India are a group of very feminine men who worship the Mother Goddess, Bahuchara Mata. Most of the men are homosexual. Many of the men undergo an operation in which their penis and testes are removed. This surgery is illegal and can be quite dangerous; it is performed by a member of the group, or sometimes by a cooperative surgeon. The *hijras* are paid to perform at weddings and the celebration of male births, but this service is actually more extorted by the *hijras* than solicited by families. If a family does not pay, the *hijras* make trouble, perhaps even flashing their mutilated genitalia. Many of the *hijras* also engage in male prostitution.

When he visited Tahiti, Captain Bligh (commander of the *Bounty*) noted that the *mahu* participated in the same ceremonies as women did. At first, their feminine behavior and speech led him to believe that they were castrated, but he learned otherwise. He observed with disgust the practice of men rubbing their penises between the *mahus'* thighs. The contemporary *mahu* fellate the men they have sex with, who do not return the favor.

The *xanith* perform women's chores in highly sex-segregated Oman and are classed with women for many social purposes. Their clothing and physical presentation is a mixture of male and female, perhaps because they are denied by law the right to dress as women. Their attractiveness is judged by female standards of beauty (white skin, large eyes, and full cheeks, for example), and they serve as homosexual prostitutes.

Evidence for the cross-cultural ubiquity of transsexuals comes from newspapers as well. Thailand holds an annual transsexual beauty pageant. (The most recent winner, 22-year-old Thanaporn Wongprasert of Bangkok, said she would spend the $1,300 prize money to make herself more beautiful.) A recent article in the *Bangkok Post* discussed Thai television's excess of drag queens: "Tune into any daytime television channel these days, and before long you may well see a character who is male, but whose behavior is a wild parody of female mannerisms: exaggerated ladylike daintiness, drag-queen hijinx, campy verbal acrobatics—they're all in the repertoire. No woman could hope to compete with their version of over-the-top super-femininity. And evidently audiences love it. Male cross-dressers are now familiar figures, guaranteed to bring an instant laugh in TV comedies, dramas, talk and game shows, commercials, and even films."

From Tonga: Thirteen contestants entered the annual Miss Galaxy transgendered beauty pageant July 9-11 in Nuku'alofa, capital of the South Pacific nation of Tonga. In the end, Natasha Pressland, 18, emerged victorious. Her hobbies are "dancing, praying, and meeting people." She plans to become a flight attendant.

The sociologist Fred Whitam has spent much of his career studying homosexuality in non-Western cultures. Based on his observations in Brazil, Guatemala, the Philippines, Indonesia, Thailand, and the United States, Whitam is convinced that drag queens and transgender homosexuals exist in all societies. Furthermore, Whitam has found that within homosexual communities, the more important distinction is between conventionally masculine gay men and the very feminine drag queens and transsexuals. Despite particular cross-cultural wrinkles, Whitam has recognized several cross-culturally universal characteristics among the latter. Invariably, drag queens and transsexuals were highly feminine boys who exhibited low levels of athletic interest and high levels of interest in activities considered feminine. As adults, they use female pronouns to talk about themselves, and they adopt female names. They tend to dress in a theatrical, glamorous manner. All drag queen and transsexual communities of any size produce conspicuous entertainment forms, and the most appealing occupations to the members are entertainment-related (primarily singing, dancing, and acting). Transgender homosexual men often work in occupations that are viewed as traditionally female or "gay," such as prostitution, hair styling, sewing, housekeeping, or manicuring. They have high levels of interest in having sex with men, and their partners tend to be heterosexual or bisexual, rather than gay men.

Whitam believes that one of the most culturally variable phenomena is the willingness of straight men to have sex with very feminine gay men. In America, this appears to be a rare practice. However, in some other cultures Whitam says it is common. For example, in the Philippines many straight adolescent males have their first sexual contact with *bayot*, or members of the transgendered gay male tradition there. Sexual liaisons with *bayot* are thought of as adolescent peccadilloes no worse than smoking and drinking. They are certainly more acceptable, in certain respects, than spoiling the virginity of "nice girls."

Whitam's observations are necessarily limited to contemporary societies, although it is unlikely that any of the transsexual and drag queen subcultures has arisen recently (due, for example, to the international televising of daytime talk shows). The cross-cultural regularity of homosexual transsexuals and drag queens is highly suggestive of some fundamental biological influence that transcends culture.

★★★★★★★★

Egalitarian homosexuality is the type of homosexual relationship most common in the contemporary West. Egalitarian homosexuality occurs when two individuals of similar age and class form a homosexual relationship. For some reason, during the second half of the twentieth century, this has become the predominant form of homosexuality in the West, while transgender homosexuality has become rare. However, these two forms of homosexuality are not as different as they appear.

Take Danny Ryan. Chances are that when he grows up he will be a gay man practicing egalitarian homosexuality. There is a smaller chance that he will become a woman, and if he does, his sex life will be of the transgender type. If Danny had been born among the Crow Native Americans during the 1800s, he would almost certainly have been made a member of the *berdache*. Egalitarian and transgender homosexuality are similar because they contain some similar people. Transgender homosexuality is rare in Western culture, but this is not because few men have the potential to be transgender homosexuals under the right circumstances. My research, which demonstrates a large degree of femininity among gay men, suggests rather that the Western gay community has plenty of men who would have been candidates for transgender homosexuality in cultures where this was the main type of homosexuality.

The main difference between transgender and egalitarian homosexuality is that in transgender homosexuality, only one of the partners is truly gay. The partner in the feminine role is gay because "she" wants men. Furthermore "she" wants men who are masculine. What about these masculine men? Although they are in some sense engaged in homosexual behavior, because they are having sex with someone born a man, they do not view the transgendered partner as a man. Some view "her" as a woman; others as a member of a third gender, neither male nor female. The masculine partners prefer either women or males imitating women.

In contrast, both the members of an egalitarian homosexual relationship are gay. Both want masculine men. Neither partner in this kind of relationship would be excited to have a male partner pretending to be a female. Even if one of the partners would actually enjoy taking the feminine role—taking a female name, wearing a dress, pretending to be a woman, basically becoming a full-time drag queen—he would refrain in order to make his partner happy. If a gay man wants to attract straight men, he should imitate a woman. If he wants to attract gay men, he must stay a man.

I think it is likely that many of the men on Halsted Street could have been members of the *berdache* or the *hijras*. Some may actually have been happier if they had been born into a society in which the transgendered role was more common. If Danny Ryan had been born into such a society, he would not have to throw off or hide his femininity. It would be a necessary part of his role. Some cultures with a prominent tradition of transgender homosexuality even help feminine boys into the transgendered role. But in the contemporary West, this doesn't happen. Here, Danny must learn to act more like a boy, and he must become a man.

Despite these pressures, a few males resist. They are reluctant to give up their ambition to become women, and they decide to pursue their dream. The rest of the book is about them.

Part III

Women Who Once Were Boys

*I*t is 2 AM Sunday night (actually Monday morning) at Crobar, and I am tired. I have had only limited success tonight recruiting research subjects for our study of drag queens and transsexuals and am cruising the huge club one more time before leaving. The Crobar crowd is on a different schedule than I; the place is just reaching its peak intensity. I pass a tall, attractive, black woman, who sees me staring at her, and somehow she understands what's on my mind. "No, I'm real," she laughs good-naturedly. (I am thankful that none of the dozen or so genetic females we mistakenly approached during the course of this study ever became hurt or angry. I wonder if they understood that the implication that we thought they might be transsexual was not an insult. Many of the transsexuals we interviewed in the course of the study were more attractive than the average genetic female.)

I start upstairs to get the panoramic view and I see Kim for the first time, on the stairs, dancing, posing. She is spectacular, exotic (I find out later that she is from Belize), and sexy. Her body is incredibly curvaceous, which is a clue that it may not be natural. And I notice a very subtle and not-unattractive angularity of the face, which is also not clearly diagnostic on this tall siren. It is difficult to avoid viewing Kim from two perspectives: as a researcher but also as a single, heterosexual man. As I contemplate approaching her, I am influenced by considerations from each perspective. I have this strong intuition that I am correct about her, but if I am not, I might have the unpleasant experience of simultaneously insulting, and being rejected by, a beautiful woman.

As I waver, I notice her companion, an attractive, blonde-haired, blue-eyed man whose body, amply displayed in a tight tank top, is the male analogue of Kim's—he has a huge chest (hairless of course) and bulging biceps. They are a beautiful couple, or at least a couple of beautiful people. They dance together, occasionally smiling at each other, but they do not dance closely or in a way that betrays the sexual desire that virtually anyone would feel toward at least one of them. I remind myself that it is Sunday night "Glee Club"—gay night at Crobar. What would a gorgeous heterosexual couple be doing at Crobar tonight at 2 AM? In fact, however, this is a very trendy setting even among heterosexuals, particularly if one is unconventional and open-minded. (Dennis Rodman has been a regular.) I cannot decide whether Kim is transsexual, and in a tribute to her beauty, I decide for now not to approach her. If she is transsexual, I will have other chances to meet her, and I will probably also have the opportunity to find out from others without asking her directly. So I leave.

★★★★★★★★

Based on the frequency of their appearance on American talk shows—"Beautiful Women Who Used to be Men," "My Wife Used to be a Man," "My Husband Is a Woman," "My Husband Has Become a Woman"—transsexuals might appear to constitute a sizeable minority. They do not. Fewer than 1 in 20,000 persons is transsexual. Most of us do not personally know a transsexual, although many of us have had the experience of wondering if a particular woman we have seen is actually a man, and most of us who have been to even a few gay bars have seen one. There are also transsexuals who work as waitresses, hairdressers, receptionists, strippers, and prostitutes, as well as in many other occupations, whom we might meet incidentally, without even questioning whether they might have once lived as men.

Transsexuals appear frequently on daytime talk shows not because they are common, but because people find them fascinating, and because talk shows' continued existence depends on their catering to people's fascinations, no matter how elevating (or not) those may be. I have been asked to talk to respectable media as an expert on transsexualism regarding two cases: a race car driver who got a sex change, and a Chicago area teacher who was living as a man in the spring and assumed a female identity in the fall. *Time* magazine recently ran a story about the "transgendered," and movies in recent years such as *Priscilla: Queen of the Desert, The Crying Game, Ed Wood, Silence of the*

Lambs, and *Midnight in the Garden of Good and Evil*, among others, have featured transsexuals and their like. Transsexuals are hot. I have discerned a few main themes in transsexual mania, which include:

"What is it like to feel that you were born the wrong sex?"

"What would it be like to become involved, knowingly or not, in a romantic relationship with a transsexual?"

"Isn't it amazing how convincing a woman she makes?

Alternatively (but not in the same show),

"Isn't it strange to see that male-looking person proclaiming his inner femininity, dressed like a woman, and evidently somewhere in the process of obtaining a sex change, when he used to be married and work in a bank?"

"What is involved in getting a sex change?"

"Do transsexuals ever regret their surgery?"

These are, indeed, all fascinating questions. And if they are less practically important than the question of how to reduce the national debt, some issues related to transsexualism do touch on fundamental issues about human nature. Unfortunately, the typical format in which these are discussed is designed to provoke rather than to illuminate. Moreover, it is not surprising that the typical television viewer has only superficial knowledge about transsexuals when many "experts" who make their livings working with them do not understand transsexuals very well.

Most people—even those who have never met a transsexual—know the standard story of men who want to be women:"Since I can remember, I have always felt as if I were a member of the other sex. I have felt like a freak with this body and detest my penis. I must get sex reassignment surgery (a "sex change operation") in order to match my external body with my internal mind." But the truth is much more interesting than the standard story.

****

Terminology is an important source of confusion. "Transsexualism" has many connotations, including "sex change," "trapped in the wrong body," "femininity" (in genetic males) or "masculinity" (in genetic females), and "cross-dressing." Try to forget the connotations. All I mean by "transsexualism" is the desire to become a member of the opposite sex. An adult with transsexualism is a "transsexual." These definitions say nothing about the motivation, appearance, or subsequent actions of the transsexual. They do not imply that the transsexual feels trapped in the wrong body, or that the transsexual even ultimately seeks sex reassignment.

The definitions also allow different degrees of transsexualism. I have no desire to become a woman, and so I am not at all transsexual. But there is a range of transsexualism among people who do have such desires. There are people who would like to change sex if they could try it out for a while and change back if they chose. (I do not mean to include people who have whimsical thoughts about this. Only serious thoughts.) Some would change sex permanently if they did not have to give up important aspects of their lives (for example, their children) to do so. Others have no ambivalence, and might dedicate their lives to changing their sex to the point of apparent obsession, losing families, friends, and jobs in the process. All these people I have described are at least a little transsexual, but the latter are more transsexual than the former. "Transsexual" is not an either-or label. Even if we were to restrict the use of that word to those who take medical steps to change their sex, there would still be considerable variability. Some transsexuals merely undergo electrolysis; others take hormones; others get breast implants; and of course, others get an operation to simulate the genitalia of the other sex.

Despite this variability, it is possible to get a handle on the psychology of male-to-female transsexualism. You just have to meet the right people—starting with some transsexuals, themselves.

CHAPTER 8

Terese and Cher

One way that the standard transsexual story is wrong is in its singularity. Two different types of men change their sex. To anyone who examines them closely, they are quite dissimilar, in their histories, their motivations, their degree of femininity, their demographics, and even the way they look. We know little about the causes of either type of transsexualism (though we have some good hunches about one type). But I am certain that when we finally do understand, the causes of the two types will be completely different.

To anyone who has seen members of both types and who has learned to ask the right kinds of questions, it is easy to tell them apart. Yet the difference has eluded virtually everyone who cares about transsexuals: talk show hosts, journalists, most people who evaluate and treat them, and even most academics who have studied them. One reason is that the superficial similarity of the two types is so striking—both are men, usually dressed and attempting to act like women, who want to replace their penises with vaginas—that it prevents us from noticing more subtle, though also more fundamental, differences. Another reason is that the two types of transsexuals rarely show up side by side, where they would be easily distinguishable. In the United States in the third millennium, they do not use the same "gender clinics," and although they often associate with other transsexuals, this is nearly always with their own type. The most interesting reason why most people do not realize that there are two types of transsexuals is that members of one type sometimes misrepresent themselves as members of the other. I will get more specific later, but for now, it is enough to say that they are often silent about their true motivation and instead tell stories about themselves that are misleading and, in important respects, false.

The two types of transsexuals who begin life as males are called *homosexual* and *autogynephilic*. Once understood, these names are appropriate. Succinctly put, homosexual male-to-female transsexuals are extremely feminine gay men, and autogynephilic transsexuals are men erotically obsessed with the image of themselves as women. When most people hear "transsexual" they think of the homosexual transsexual, who fits the classic pattern. From soon after birth, the homosexual male-to-female transsexual behaves and feels like a girl. Unlike most feminine boys (such as Danny, from Chapter 3), these transsexuals do not outgrow, or learn to hide, their femininity. Instead, they decide that the drastic step of changing their sex is preferable. They unambiguously desire and love men, especially heterosexual men, whom they can attract only as women. (I recognize that using the term "homosexual" to describe a type of transsexual is confusing. Transsexualism terminology is often confusing. I often have to think hard about whether to write "he" or "she," for example. In this case, there is scientific precedent—"homosexual transsexualism" was proposed by the scientist who first discovered that there were two types of transsexuals. The term also is conceptually revealing, because one type of transsexual man is a kind of homosexual man. Read on; you'll get used to it.)

Honest and open autogynephilic transsexuals reveal a much different pattern. They were not especially feminine boys. The first overt manifestation of what led to their transsexualism was typically during early adolescence, when they secretly dressed in their mothers' or sisters' lingerie, looked at themselves in the mirror, and masturbated. This activity continued into adulthood, and sexual fantasies became increasingly transsexual—especially the fantasy of having a vulva, perhaps being penetrated by a penis. Autogynephilic transsexuals might declare attraction to women or men, to both, or to neither. But their primary attraction is to the women that they would become.

These summaries are given here less to clarify than to organize the information that follows. Transsexualism—especially autogynephilic transsexualism—seems so foreign to most people that it requires a great deal of explanation. And illustration.

So meet Cher and Terese, most extraordinary friends. Terese is shy, small, and very feminine. Cher is extra extroverted, tall, and in some ways strikingly masculine. Terese is Mexican-American, Cher Italian-American. Both Terese and Cher are transsexuals born boys, now women. The most unusual fact about them in my experience is that they are close friends *and* are different types of transsexuals; serious socializing between the two types is practically unheard of. Terese is a homosexual transsexual, and Cher is autogynephilic. Spend the day with them, listening to their stories and watching the way they behave, and the difference between homosexual and autogynephilic transsexualism will be forever etched in your mind.

★★★★★★★★

Terese was born Jose Garcia, in Mexico. His parents divorced when Jose was young, and he was raised by grandparents. They moved to Chicago when he was 10. From early childhood, Jose wanted to be a girl. "Why do girls get prettier things and why do they get to do the things I like to do?" he wondered. He knew, because everyone around him told him, that he was a boy. But that did not guide his behavior. He dressed up as a female as often as he could, played with dolls, was the "mommy" when playing house, and avoided the rough boy sports. His best friends were girls. (In all these senses, he was just like Danny, whom we met earlier. Danny probably will not become transsexual, though he will probably become a gay man. Why most very feminine boys grow up to be gay men and a few get sex changes is not known, though I'll speculate about it later.)

Although his friends were mainly girls, Jose was also intensely interested in boys, especially when he got to junior high. He began having crushes on them, especially the athletes, and even some of his male teachers. He never experienced sexual feelings for a girl or woman. When he was 13, an older boy initiated Jose into sex. Jose is certain that the older boy is now a gay man. He had several gay partners—Terese calls this her "gay boy" phase—but began to notice that he did not find other gay boys or men to be particularly attractive. They were too feminine. Jose was much more attracted to the straight men that he met. Also, Jose did not like it when others touched his penis (or even when he touched it himself). He preferred giving others oral sex, or being penetrated anally, though this latter activity frequently hurt.

When he was 14, Jose found some gay friends with similar inclinations toward femininity. They began going out to parties, or shopping, in drag. It was fun to get away with cross-dressing. Jose was a late maturer, with a smooth complexion and no facial hair, so he passed well. And the straight men he met when in drag frequently pursued him. This was exciting but frustrating, because obviously, Jose could not let them complete their advances.

Jose was lucky to be well liked at school so he was not teased as mercilessly as many very feminine boys. But his family was not happy with his development. His father met him after many years and told him to "stop acting like a girl." His mother was also disturbed and worried that he would be gay. As he got older, he made more effort to appear masculine, though he kept his hair long. Pictures from this time show Jose to be a rather androgynous young man, but clearly a man.

He did not feel more masculine, but merely put his femininity in the closet.

Jose took a job at the Chicago Board of Options that required him to dress conservatively. During this time, he became depressed. He dreaded waking up and putting on a tie every day, and hiding his femininity was sufficiently stressful that his hair started falling out. One day at age 19, Jose was brooding about his future. His voice was finally deepening noticeably and he was getting facial hair. He decided that he could not live the rest of his life as a man, even if it meant being abandoned by his family. The next day, Jose began to live full time as a woman, and Terese was born. When she told her friends and family of her decision, her mother expressed relief rather than shock. She had known from early on that Terese's identity was more feminine than masculine. Only one friend, a gay man, objected to Terese's decision. He said that she was "selling out," abandoning a gay identity because she could not deal with social ostracism. (Terese disagrees that this played any part in her decision.)

Terese had learned over the years where homosexual transsexuals congregated (primarily gay bars and dance clubs as well as a few small bars that featured female impersonators), and she increasingly frequented those places. She felt more comfortable there, and she discovered that she could benefit from the cumulative knowledge of "girls" in her situation. For example, an important immediate consequence of her transition was her decision to begin taking the female hormone, estrogen, to prevent her body from becoming more masculine, and to feminize it. Other transsexuals told her where she could get estrogen on the black market and how much to take. (This was risky because estrogen can have serious side effects, although the worst ones, such as blood clots, are rare. In Chicago these days, though, most homosexual transsexuals do not bother with official medicine because physicians charge so much and because they are used to living beyond society's rules.) She also learned about sex reassignment surgery, surgery to change the penis into a neo-vagina: how much surgery cost at different places and which surgeons did the best job.

Terese had several sexual relationships during this time. In each case, the man assumed she was a non-transsexual female, and Terese did not initially tell him otherwise. She would interact sexually with him, having oral sex for instance. Only after she saw him several times, both of them growing more attached, and the issue of intercourse arose, would she consider telling him the truth. Twice, she risked honesty. Both times were traumatic, the men disappointed, angry, and repulsed, Terese hurt and afraid. Still, one of the men got used to the idea, and they continued romantically and sexually, for a while. During sex, Terese always behaved as a woman, and this partner never touched her penis. He continued to think of Terese as a woman, albeit an unusual one. Still, the relationship did not go anywhere, and the man did not treat Terese very well. She discovered that he had a steady girlfriend, for example. Although many naturally born women have been mistreated by men, Terese thinks the fact that she was a she-male (a woman with a penis) prevented men from committing to her.

Terese lived as a woman for three years before she got enough money together (about $10,000) to get her operation. Part of the money was a loan from Cher, who had become one of her best friends. In July 1997, Terese (then 25) flew to Belgium and over a four-day period, had sex reassignment surgery, learned to care for her new vagina, and recovered sufficiently to leave the hospital. Within three months, her neo-vagina had healed, and she lost her neo-virginity soon after.

In many ways Terese has blossomed since her surgery. She looks great. Not only do people fail to notice that she is a transsexual, but most men find her sexy and attractive. Depressed and in self-imposed isolation when I first saw her, she is flirtatious, energetic, and socially busy now. Among other things, she models lingerie. She has dated and had sex with several heterosexual men, none of whom knew about her past life. (She is still looking for a serious boyfriend.) Her mother is thrilled to have her new daughter. For one, she is relieved for Terese, knowing how much she wanted this. But she is also happy to have a very attractive feminine daughter rather than an unhappy and noticeably feminine son. When she visited Terese, she proudly took her new daughter shopping.

In an attempt to make a new life for herself, Terese has distanced herself from her old transsexual friends. When she goes out to meet men, she does not want to have her secret revealed by being with people who are visually identifiable as transsexual, or who are widely known to be transsexual. This has caused varying degrees of difficulties with her old friends. Cher, for one, is deeply hurt, and they have fallen out as friends. However, even Cher admits that Terese's new life appears to suit her.

In important respects, Terese's story is the story of all homosexual male-to-female transsexuals. Her early, extreme, and effortless femininity, her unambiguous preference for heterosexual men as sex partners, her (however brief) attempt to live as a gay man, and her difficulty in securing the right kind of guy prior to surgery, are almost universal among this type of transsexual. There are some differences among them, to be sure. For example, although Terese disliked her penis, some homosexual transsexuals not only like, but also use, their penises while they have them. Some, like Terese, alternate between unsatisfying and limited sexual relationships and self-imposed isolation; others earn good livings as she-male prostitutes (more on this later). These differences reflect differences in personality styles. Fundamentally, all homosexual transsexuals are similar, and after a bit of experience, easily distinguishable from the other type of transsexual.

★★★★★★★★

Cher, né Chuck Mondavi, was born in Chicago to lower middle-class parents. Cher remembers her father as a nervous, punitive, and eccentric man, who had a nervous breakdown when Chuck was young. Chuck was sometimes withdrawn and shy, but not feminine. He participated in boys' sports and activities and had male friends. Cher remembers that Chuck had occasional wishes to be a girl as early as age six. However, no one but Chuck knew of these, and no one else would have surmised this based on Chuck's behavior.

In late childhood or early adolescence, around the age of 12, Chuck was an altar boy at early mass. His parents went to mass later, giving Chuck free rein in the house. During this time, he began to dress secretly in his mother's lingerie: panties, girdle, or bra. He would look at himself in the mirror, become increasingly sexually aroused, and then masturbate to orgasm. Sometimes when his parents were home, he would wear lingerie under his male clothing, look at the lingerie models in a Sears catalogue, and rub his penis against the bed, making sure to stop the activity when his mother entered the room. While looking at the catalogue, he imagined that he was the lingerie model. Chuck was artistic, and one early creation involved painting some coconut shells flesh color, and wearing them as breasts during cross-dressing.

As his adolescence progressed, Chuck cross-dressed and masturbated at least once per week. He also masturbated while looking at *Playboy*, and his fantasies alternated between having sex with the centerfold and recreating her with his own body. As a senior in high school, he worked in a store that carried women's clothing. One night he worked late, mopping floors, and as he noticed the hosiery and fishnet stockings, could not resist the urge to take some home to use in his cross-dressing sessions.

Chuck attended an all-boys high school. Although Chuck might have been considered a bit eccentric, his classmates did not suspect that he was a cross-dresser. Cross-dressing was his secret life. He believed that his activities were unusual, and he was often ashamed of them, so when he went to college, Chuck tried to "purge" by not bringing clothes along. When he participated in a panty raid, and acquired several bras and pairs of panties, he hung them up in his room as trophies. When his roommate left their dorm room, Chuck gave in to his urge to cross-dress.

His artistic creativity was flourishing. He won a painting competition and the art faculty began to buy his paintings. He kept quirky hours, working at all hours of the night at a studio, where he could usually cross-dress in private. He ultimately dropped out of college because he thought he could make it as an artist. Cher refers to what followed as Chuck's "bum period." During this time, Chuck began feeling depressed about cross-dressing because it made him feel "freakish" and "lonely." (Chuck was still a virgin, and did not have an active dating life.) He sought behavior modification therapy and stopped cross-dressing for about five years. During this time, he met and confided in a psychologist, who, Cher believes, wanted to cure Chuck, and they became romantically involved. At age 33, Chuck lost his virginity. Chuck did not find the relationship gratifying, however, and it did not last.

After the break-up, Chuck was driving through a poor neighborhood and noticed a wig in a garbage can. He stopped the car and waited, struggling with the intense urge to get out and get the wig. He drove on, but after a few blocks the urge was so intense that he drove back to the garbage can and collected the wig. Oddly, there were other items of women's clothing with the wig—Cher suspects that Chuck had stumbled on the results of a cross-dresser's recent purge. Soon after this, Chuck made peace with cross-dressing and stopped worrying whether it was normal or acceptable.

This began a period in Chuck's life marked by a devotion to cross-dressing that was both obsessive and highly creative. He began collecting pornography that featured pictorial stories of women in various stages of dress and undress (nurses, stewardesses, and so on). He searched junk clothing stores and clothing catalogues to recreate their costumes, and cross-dressed and fantasized about them in private.

Chuck's fantasies began to involve more than looking like a woman. Increasingly, he began to fantasize about having a vagina, and about being penetrated by a man. He was not attracted to any specific man, and he did not fantasize about specific men, only faceless men, and their penises. He enjoyed the fantasy of being physically dominated, as well as vaginally penetrated, by a man.

His activities also progressed. He used a coke bottle to penetrate his anus, fantasizing that it was a penis penetrating his vagina. Chuck's artistic talents began to come into play. He constructed a "robot man" that could fulfill the fantasy of penetration. "Robot man" had a body, a penis made of a dildo, and even an arm that Chuck could manipulate to make it feel as if it was stroking his back. Chuck attached a mirror to his bedroom ceiling, and could view the image of the robot man on top of Chuck, dressed as a woman, "penis" in Chuck's anus.

Chuck also began to take more elaborate steps to create the image of himself as a woman. He wore fake breasts and a feminine mask. (Chuck still had a beard.) He purchased several fake vaginas from adult bookstores. Although these are made for heterosexual men to view and penetrate, Chuck reversed that function. Born with one testicle, he discovered that he could invert his penis and scrotum into his body. He then glued a fake vagina over his inverted genitals. Ever the artist (and an exhibitionist as well), Chuck frequently videotaped himself as a woman. In one segment, Chuck begins by standing "naked" (except for shoes) with mask and fake breasts and fake vagina. As Donna Summer sings "Love to Love You Baby" in the background, Chuck begins to walk toward the camera, somewhat awkwardly in high heels. Cut to Chuck, similarly "dressed," astride a dildo (which is anally penetrating him, though the image is constructed to look as if the penetration is vaginal), with the soundtrack (from a porno film) of a woman's sexual moaning.

These activities, and the fantasies they fulfilled, were enormously erotic to Chuck, and they were all-consuming. He spent all night cross-dressing and enacting fantasies, and slept all day. He had abandoned art (except for his artistic contributions to his sexual fantasies). His family was concerned, because they felt that he was physically neglecting himself. He did not bathe regularly, for example, and others complained that he stank. Most of Chuck's physical energies were devoted to cross-dressing and related sexual activity. Cher recalls that once when Chuck was cross-dressed and having sex with "robot man," the ceiling mirror partially detached and hit him in the head. He wondered what people would have thought if he had been killed and his body subsequently discovered. He began to doubt his sanity, but could not stop. He was becoming seriously depressed. He was lonely, and disgusted with what he felt was his own narcissism.

Soon after this, he sought treatment for depression. He confided in a psychiatrist about his secret life. By coincidence, the psychiatrist had worked with John Money, the eminent sex researcher. He suggested that Chuck might be a transsexual rather than "merely" a transvestite (or heterosexual cross-dresser). Subsequently, Chuck was assessed at a gender clinic in Wisconsin, where he was diagnosed a transsexual. This was an enormous relief and he felt that his obsessions were now more explicable. Chuck now knew what he wanted to do with his life: become a woman and have sex with other people rather than by himself. He dismantled "robot man" and began to plan for his transition.

Cher (named for the performer, whom Cher feels she resembles) was born in 1991. Her mother had died by then, but her dying father accepted her decision in good spirits. His death later that year gave Cher the inheritance she needed to pay for her surgery. One year and a day after Chuck became Cher, Cher got sex reassignment surgery in Montreal. She was 40 years old. By then, both parents were dead, and many of her remaining family rejected her. (This was due in part to disagreement about the disposition of her father's estate.) Despite this, for the most part Cher has been happier than Chuck was. She is more outgoing and feels that she lives a real life now, instead of a fantasy life. Despite her negative experiences with her family, many other people have accepted her. For example, she attended the 25th reunion of her all male high school class—as a woman. Pictures of that event clearly show that her classmates enjoyed and admired her. She continues to play the dulcimer in the Irish folk music group she helped form as a man; her fellow musicians have no problem with this.

During the time between Cher's birth and her full transition, she had sex with several men, always involving their penetrating her anally. Always, in these experiences she thought of herself as a woman rather than a man. Although she remained attracted to women, she did not manage to have sex with any. Cher insists that once Chuck became Cher, the sexual focus was no longer a self-image, but other people. After her sex change, she had vaginal sex with several men. Nevertheless, she complains about her sex life and has not had a partner lately. Many men can tell that she is transsexual, and Cher insists on being honest about her past before becoming sexually involved. The fact that Cher used to be Chuck is a problem for most men. Some men are simply rude and cruel to her. Unlike many of her friends, Cher is unwilling to engage in casual sex with men (or women) and is waiting for someone who wants to spend time with her and date her. When she gets sexually aroused, she still masturbates simulating intercourse with a dildo while looking in a mirror; now the dildo penetrates her vagina rather than her anus.

I have never met anyone quite like Cher before. This is in part because Cher is a character, and she would be a character even if it were not for her unusual sex life and sex change. She is eccentric, talkative, and exhibitionistic. (One of the most strangely hilarious experiences I have ever witnessed was Cher lecturing to a group of gifted adolescents about transsexualism. She began dressed as a rough-talking man and eventually stripped down to Cher, in a bikini. I will never forget the wide-open, 16-year-old mouths.) Although some elements of Cher's story are very common to this kind of transsexual (especially the erotic cross-dressing), others (such as the wearing of fake vaginas) are unique to her. At least I have never met other transsexuals who admitted to this. Nevertheless, I think that Cher is a wonderful example of the second kind of transsexualism, less because she is representative than because she openly and floridly exemplifies the essential feature of this type, which is autogynephilia.

CHAPTER 9

Men Trapped in Men's Bodies

ay Blanchard, Head of the Clinical Sexology Program of the Clarke Institute of Psychiatry in Toronto, knows more about transsexualism than just about anyone else. Like so many sex researchers, Blanchard's introduction to sex research was fortuitous rather than intentional. A rat psychologist by training, his first job was as a clinical psychologist at the Ontario Correctional Institute. He worked full time evaluating and treating inmates and was miserable. "I didn't want to spend 100 percent of my time as a front-line clinician. I wanted to make a scientific contribution. Plus, the prison environment was stressful, if never dull." During that time, the eminent sex researcher, Kurt Freund, consulted at the hospital. Someone suggested that they meet, and during their first conversation, they made plans to collaborate. In 1980 Blanchard took a job at the Clarke, where he has remained, recently taking Freund's position after his death.

Blanchard is irreverent, cynical, and politically incorrect. During the opening ceremony of the International Academy of Sex Research, during the eulogies for members who died during the previous year, he regularly engages in wickedly entertaining whispered commentary about the deceased, unsentimentally recalled. (His eulogy for Freund, however, was serious and touching.) He has little patience for arguments about whether research is good for people (such as "Are homosexual people hurt or harmed by research on the genetics of sexual orientation?"), preferring instead to dwell on whether scientific findings are true. A transsexual colleague of Blanchard's tells how she once sought his advice regarding a professional dilemma. A member of a task force about transsexualism, she disliked the first draft of their report, but was worried she would offend the other (non-transsexual) members of the committee. Blanchard's characteristic advice: "What's the point of being a sacred cow if you don't moo?"

Academics remember each other by their "contributions," or ideas that make a mark, ideas that get the attention of other academics, motivating others to study the same thing. Blanchard has made two contributions so far. His most recent work has shown convincingly that gay men tend to have more older brothers compared with heterosexual men, and he is pursuing an interesting biological theory to account for this. But I think his more revolutionary contribution has been to the classification and explanation of transsexualism. In a series of articles beginning in 1985 and continuing for about a decade, Blanchard established that there are two very different types of men who change into women, who have very different presentations, motivations, and probably, causes. Blanchard's observations transformed male-to-female transsexualism from a seemingly chaotic and bizarre collection of phenomena into two straightforward and clinically comprehensible patterns.

****

When Ray Blanchard began studying and treating transsexuals

during the early 1980s, the field was rife with different confusing diagnostic typologies, including anywhere from one to four kinds of transsexuals. At a merely descriptive level, most clinicians saw a similar array of gender patients. First was the kind of transsexual that most of us think of when we hear "transsexual," the classic, homosexual type, like Terese. From soon after birth, these males behaved like, and desired to be, female. No one who spent much time with them could be very surprised that eventually they would choose sex reassignment surgery. Of all transsexual types, this was the one that most seemed to consist of "women trapped in men's bodies."

Another type of transsexual that specialists recognized, but which is less familiar and comprehensible to most people, was the heterosexual transsexual. These males frequently came to treatment as married men who claimed to have had hidden or suppressed longstanding wishes to be female that they could no longer deny. The clinical picture of this type was much different than the homosexual transsexual. Obviously, of course, one was attracted to men and the other to women, but there was another striking difference. Most heterosexual transsexuals, and virtually no homosexual transsexuals, reported a phase beginning during adolescence in which they secretly wore women's clothing such as lingerie, became sexually aroused, and masturbated. Cher has a history that is characteristic of many heterosexual transsexuals. However, she has had sexual fantasies about both men and women, and has had sex with both. She is bisexual.

There was at least one other type of transsexual that experts wrote about, one that presented a much fuzzier picture. This was a transsexual type whose sexuality was either unclear or absent. These transsexuals claimed to be attracted to neither women nor men, and some of them claimed to have no sexual feelings at all. Some called this type "asexual."

To make matters more complicated, there were at least two other types of men who cross-dressed, but who did not want sex changes. Drag queens were gay men who cross-dressed in public, often for the purpose of entertainment (especially lip synching and dancing to music). Transsexualism researchers knew that drag queens existed but did not know much about them because drag queens did not often seek treatment for their cross-dressing; it caused them no problems. The other group of men who cross-dressed were called transvestites. These were heterosexual men, typically married, who dressed in women's clothing without any intention of getting a sex change. Like heterosexual transsexuals, they frequently acknowledged an erotic component to their cross-dressing, at least during adolescence. The official psychiatric diagnostic manual of the time, *The Diagnostic and Statistical Manual of Mental Disorders (DSM III-R)* called heterosexual transvestism "transvestic fetishism," and put it in an entirely different section of disorders from transsexualism. The diagnostic label implied that transvestites are motivated by an erotic connection to women's clothing. Heterosexual transvestites are an important part of this story, and they are probably far more common than transsexuals (no one knows for sure). So before proceeding, let's meet one.

★★★★★★★★★

Stephanie Braverman is, despite the name, a 50-ish married man who holds a high-level position in a national bank, and whose wife is a well-known administrator in a local university. When Stephanie is a man—which is most of the time—that man is named "Don." I have never met Don. Stephanie always looks the same, dressed in tastefully elegant style—wig, makeup, and all—as a middle-aged matron. After reading about my interest in gender in a newspaper article, Stephanie called me to "educate" me about heterosexual cross-dressers. We have met several times, and I have learned much. Still, we disagree fundamentally (if good-naturedly) about many things related to cross-dressing. One thing we mostly agree on, however, is her basic story.

Stephanie remembers Don having early (age six or so), vague wishes to be a girl, but does not think that others found him remarkably feminine. In adolescence, Don periodically wore his mother's lingerie, looked in a mirror, and masturbated. He felt ashamed and guilty about cross-dressing and tried to stop several times, but each time was unable to refrain for long. In his early 20s, he met the woman he would marry. Before proposing, he told her about his cross-dressing and said that he intended to stop. His future wife, in love with him by then, did not even pause to reconsider her commitment. They have had, by most appearances, a successful marriage, with three children. Several times, during the marriage, Don "purged," throwing away all his women's clothing and vowing to quit cross-dressing. (Over all his purges, Don probably threw away several thousand dollars worth of clothing.) Following each purge, he felt relieved and virtuous for a time, but these feelings eventually transformed into irritability, tension, and despair. Each time, Don gave in to his cross-dressing urges. After his last (and final, Stephanie hopes) purge attempt failed, he sought counseling, and his therapist helped him find the Chicago chapter of the Society for the Second Self (or Tri-Ess), an organization for heterosexual cross-dressers. Stephanie has become a leader in the chapter and speaks frequently (and always as Stephanie) to college students and other receptive audiences. She wants ultimately to diminish the stigma felt by cross-dressers.

Stephanie claims that the erotic component of cross-dressing was never Don's primary motivation—which was the expression of an inner femininity—and that he rarely feels aroused any more when cross-dressed. Once in a while, a glimpse in the mirror of how good Don looks as Stephanie (especially his legs) gives him an erection, but this is more annoying than satisfying. Stephanie insists that the primary benefit of cross-dressing these days is relaxation. When business and family duties prevent Don from cross-dressing for much longer than two weeks, he feels tense and irritable. To me it sounds as if Don's cross-dressing was at one time primarily sexually motivated, and indeed I suspect it still is to a large extent. This is our ongoing argument and I will return to it.

****

Diagnosis can advance by either splitting or lumping. Splitting occurs when someone notices two or more superficially similar conditions that had previously been called the same thing. The history of medical diagnosis has mostly been a story about splitting. Some early classification systems, for example, included the category "fever," which subsumed every problem that caused a temperature. In contrast, modern diagnosis distinguishes thousands of conditions that cause fever, from chicken pox to Ebola.

More rarely, diagnosis progresses by lumping. Lumping occurs when two or more apparently different conditions are found to be different forms of the same, underlying pathology. In the early history of AIDS, doctors noticed an increase in a variety of seemingly unrelated conditions, such as pneumocystitis carinii pneumonia, Kaposi's sarcoma, and cytomegalovirus retinitis, among gay men. Eventually, they hypothesized that these were all consequences of the same infection, and when it was possible to test for HIV, its presence was confirmed in virtually all cases.

Ray Blanchard's contribution to transsexual science was of the lumping variety. Distinguishing "homosexual," "heterosexual," "bisexual," and "asexual" transsexuals diagnostically makes sense only if the different types have fundamentally different causes. Otherwise, why not distinguish "tall," "medium-sized," and "short" transsexuals, or "blonde" and "brunette" subtypes? Blanchard noticed some similarities between some of the subtypes that made him suspect that they were fundamentally similar. In particular, the homosexual subtype seemed different from all the others, who seemed similar in important respects. (In order to avoid wordiness, he called the heterosexual, bisexual, and asexual transsexuals the "nonhomosexual" transsexuals.)

For one, the nonhomosexual types were older when they came to the Clarke Institute for treatment. This was partly due to the fact that many of them had postponed their gender concerns to raise families.

But even the asexual subtype, who was typically unmarried, came to treatment later than the homosexual subtype. On average, the nonhomosexual types showed up at the Clarke eight years later than the homosexual type. The nonhomosexual types also gave somewhat different childhood histories than the homosexual type, who universally acknowledged marked and noticeable femininity from early on. In contrast, many of the nonhomosexual types said that they were unremarkably masculine boys, though they typically reported early memories of secret longing to be girls.

The most noteworthy difference between the homosexual and nonhomosexual transsexuals concerned cross-dressing. Homosexual transsexuals recalled that they cross-dressed from early on, but few of them reported that cross-dressing provoked a sexual response. Most nonhomosexual transsexuals admitted sexual arousal to cross-dressing, at least in the past. Even the asexual subtype did so.

Blanchard made a good case that "heterosexual," "bisexual," and "asexual" transsexuals were more like each other than like homosexual transsexuals, and he suspected that they were subtypes of a more general condition. But what general condition? Then Blanchard saw a patient named Philip, who proved to be the exception that revealed the rule.

Philip was a 38-year-old, unmarried, business executive who was referred to the Clarke Institute of Psychiatry because of the persistent wish to be a woman, accompanied by depression that he was not one. Philip was a happy, masculine boy, who was popular and competent. Philip had cross-dressed only once in his life, at age six. Although he lived alone and could cross-dress easily, he simply had no desire to do so. Since puberty, he had masturbated fantasizing that he was a nude woman, lying alone in his bed. He would focus on the picture of having breasts, a vagina, and other female physical characteristics. Although Philip had had sex with several women, there were indications that this was not especially rewarding to him. In his only long-term relationship, he stopped having sex with his girlfriend after only a few months. In his last time with a woman, he had difficulty getting an erection. He has never had sex with a man. However, he had begun to fantasize about having intercourse with a man, as a woman. The man in his fantasies was a faceless abstraction rather than a real person. Philip never fantasized about having gay sex with a man, as a man.

Philip was near the borderline of nonhomosexual transsexualism, but he lacked a cardinal sign of that disorder, a history of erotic cross-dressing. It dawned on Blanchard that what Philip had in common with most nonhomosexual transsexuals, as well as cross-dressers for that matter, was sexual arousal at the idea of himself as a woman. This strange sexual desire—for oneself to be transformed into a woman—seemed to be the fundamental motivation for nonhomosexual transsexualism. Blanchard called it "autogynephilia" (pronounced Otto-guy-nuh-feel-ee-ya). "Gynephilia" means attraction to women, and "auto" means self. Autogynephilia accounts for a variety of phenomena that seem otherwise disconnected.

Take cross-dressing. At the time that Blanchard came up with autogynephilia, the prevailing explanation of erotic cross-dressing was that it was like a fetish, a mere association of sexual arousal with inanimate objects. But there were obvious problems with this conceptualization. Most men find garter belts and bras to be sexy, probably because of their association with scantily clad women, but most men do not put them on and look at themselves in the mirror. Nor do cross-dressers merely wear women's clothing. While cross-dressed, they typically pretend to be women: taking female names, trying to walk and sometimes talk like women. According to Blanchard, even cross-dressers who do not want to change their sex have autogynephilia, which they share with nonhomosexual transsexuals. This is sensible because during the teenage years, it is probably impossible to distinguish males who will become nonhomosexual transsexuals from those who will remain cross-dressers. They are all autogynephiles.

Autogynephilia also accounts for the homosexual-like fantasies of some autogynephilic (i.e., nonhomosexual) transsexuals. These fanta-

sies are quite unlike the homosexual fantasies of gay men and homosexual transsexuals. They do not focus on characteristics of the male
partner, but on the transsexual's female self interacting with the male.
Stephanie, the cross-dresser, once told me a fantasy she had been having about me. In her fantasy, I would treat her "like a lady"—take her
out to a nice restaurant and then out dancing. She reassured me that
she did not want to do anything sexual with me (and I had never
sensed that she was sexually attracted to me). As she told me about the
fantasy, however, it was clearly an erotic one. But in contrast to gay
men's fantasies about other men, I was merely a prop in Stephanie's; I
could have been any male. Even in more explicitly sexual fantasies, the
male usually has no face, just a penis and body, which penetrates the
imagined vagina. To many autogynephiles, the act of being penetrated
by a penis is the ultimate statement that one is a woman, and this is
perhaps why it is so arousing to some autogynephiles (such as Cher).

Blanchard noticed different forms of autogynephilia in the different patients he saw. Some patients were sexually aroused by cross-
dressing, others by the fantasy that they were pregnant, others by the
fantasy that they had breasts, and others by the fantasy that they had
vaginas. One patient even masturbated while fantasizing about knitting in a circle of other knitting women or being at the hairdresser's
with other women.

Blanchard hypothesized that the type of autogynephilia that a man
has should predict whether the man would become transsexual. A
cross-dresser with only transvestitic or behavioral autogynephilia can
probably satisfy his urges by periodically cross-dressing in private or in
the company of other transvestites. But a man whose primary fantasy
is having a vulva cannot enact his fantasy so easily. (Not everyone is as
creative as Cher, and eventually, wearing fake vaginas did not work for
her either.) Blanchard confirmed that, indeed, it was men who fantasized about themselves as nude women, and who focused on the image of having a vulva, who felt the strongest desire to change their sex.
He also found that patients who fantasized about themselves as nude

women were younger at their first appointment compared with those patients who fantasized about themselves wearing women's clothing. This suggests that autogynephilic transsexualism is not merely a progression from cross-dressing. If it were, the patients who fantasized about cross-dressing should have been younger, not older (because by the incorrect interpretation, they would not have yet advanced to the "nude fantasizing" stage of their condition).

Once Blanchard asked a group of autogynephilic transsexuals: "Suppose you had the following choice. You could get your sex reassignment surgery but continue to live as a man forever or live as a woman but never obtain sex reassignment surgery. Which would you do?" About half of the group chose each alternative. Those whose autogynephilia focused on the vagina chose the surgery, and those focused on the female role chose the role change. Of course, most autogynephilic transsexuals want both.

✶✶✶✶✶✶✶✶

How are we to think of autogynephilic men? Are they more like gay men, or like heterosexual men? Do they really have a woman hidden inside them? If so, why do they hide their inner femininity, when some gay men, drag queens, and homosexual transsexuals are extremely and openly feminine from an early age?

The word "autogynephilia" is difficult, even jarring, and this is appropriate—the concept it names is bizarre to most people. In order to understand autogynephilia, it is important to recognize that it differs so much from ordinary experience that it cannot be understood simply. For example, even heterosexual people can understand homosexuality by thinking, more or less accurately, "It's just as if I were attracted to my own sex instead of the other sex." Autogynephiles are more difficult to fathom.

Blanchard believes that autogynephilia is best conceived as misdirected heterosexuality. These men are heterosexual, but due to an error in the development of normal heterosexual preference, the erotic target (a woman) gets located on the inside (the self) rather than the outside. This is speculative, and what causes the developmental error is anyone's guess.

Autogynephilia is not primarily a disorder of gender identity, except in the obvious sense that the goal of the transsexual is to become the other sex. At the cross-dressers' meeting I attended, the wife of one of the men asked me: "When they say they feel like women, how do they know what that feels like?" This question, which reflected the woman's skepticism about the men's account, is profound. How do we ever know that we are like someone else? Unless you believe in extrasensory perception (and I don't), the answer must be found in overt behavior that somehow signals fundamental similarity. Evidently, the woman did not get those signals from the men. (If instead of being the wife of an autogynephile, she were the sister of a homosexual transsexual, I doubt she would have asked an analogous question.) The fact is that despite their obsession with becoming women, auto-gynephilic transsexuals are not especially feminine. One told me, for example:

> I had a fairly early onset (certainly before age six) of an intense desire to be a girl (or "like a girl") physically; or to put it another way, to be female—but not necessarily to take on the feminine gender role. This took different forms as I grew older. To a six-year-old, the difference between boys and girls physically is not primarily genital: girls wear long hair and dresses, and that's what I wanted. As I got older, I grew to want other things: breasts, a vagina, menstruation, pregnancy. In the early stages the dysphoria wasn't painful, like a wound; it was more like a yearning for the unobtainable, like a thirst I couldn't quench.

> However, I didn't play with dolls, nor do many of the traditional feminine things that the classic primaries often report. I didn't like to compete athletically, and I was always afraid of getting hurt; but I liked model cars and airplanes, and toy guns, too.

This contrasts greatly with the childhood histories of homosexual transsexuals, whose femininity was remarkable to anyone who observed them closely. Nor are autogynephiles especially feminine adults.

After all, most have been married, and to most observers, appear to be conventional husbands. Autogynephiles rarely have stereotypically female occupations. On the contrary, many have served in the military. I even met one who was in the Green Berets. Technological and scientific careers seem to me to be over-represented among autogynephiles. (Ray Blanchard remarked to me that he saw a seemingly close relation between autogynephilia and computer nerdiness.) Autogynephiles have claimed that they chose stereotypically masculine occupations to hide their feminine side, but I doubt this. It seems more consistent with the overall picture to say that autogynephilia is not associated with stereotypically feminine interests. Finally, autogynephiles do not typically look or act very feminine, especially in comparison with homosexual transsexuals. To the extent that autogynephiles achieve a feminine presentation, it is with great effort. Cross-dressers attend workshops in talking, walking, standing, and gesturing like women. The work usually pays off eventually in a passable feminine presentation, but it is work.

Autogynephiles are not "women trapped in men's bodies." (Anne Lawrence, a physician and sex researcher who is herself a postoperative transsexual, has called them "men trapped in men's bodies.") Homosexual transsexuals, so naturally feminine from early on, can make this claim more accurately, but as we shall see, it is not completely true even of them. Autogynephiles are men who have created their image of attractive women in their own bodies, an image that coexists with their original, male selves. The female self is a man-made creation. They visit the female image when they want to have sex, and some became so attached to the female image that they want it to become their one, true self. This explains the name of the transvestite organization "Society for the Second Self." It also explains the maddening tendency of some autogynephilic research subjects to put down two answers to every question—one by the female self, and one by the male self. Homosexual transsexuals do not do this. They have one self that is a mixture of masculine and feminine traits, and not alternating selves. No, autogynephiles are not women trapped in men's bodies. They are men who desperately want to become women.

****

We do not have even the beginnings of a respectable theory of the causes of autogynephilia. This differs from sexual orientation, in which we have a reasonably well articulated if unproven theory. Recall from Chapters 3 and 6 that femininity in boys and homosexuality in men are probably caused by incomplete masculinization of the brain during sexual differentiation.

Autogynephilia in the form of cross-dressing is still called "transvestic fetishism" in the *DSM IV-TR*, and many people have suggested that fetishism arises as a kind of conditioning experience. As the guy at the cross-dressers' meeting put it, panties are sexy, so some men become aroused wearing them. The problem with this account is that although it might explain a fetish for panties, it does not explain why men should want to wear the panties. Conditioning explanations of both sexual preferences and fears (the other domain where they are common) have received little convincing support, although many people believe them. Blanchard's conceptualization of cross-dressing, as arousal by an image of oneself as a woman, is very different from the idea of a fetish. Conceptualized Blanchard's way, it is difficult to see how cross-dressing could arise through conditioning.

Some autogynephiles claim that their first cross-dressing experience was in the context of being punished (usually by some female friend or relative, who forced them to cross-dress to humiliate them), and that this is how they acquired their taste for cross-dressing. I find these reports dubious. They sound more like fantasies or attempts to explain their behavior in ways that sound plausible to others. In any case, the fact that most autogynephiles do not claim these experiences suggests that they cannot be causally crucial.

Regarding the fundamental question of whether autogynephiles are born or made, my intuitions are with "born." Perhaps every day in this country, at least one adolescent boy secretly puts on his mother's or sister's lingerie for the first time, becomes sexually aroused, and masturbates. As far as anyone can tell, there is nothing unusual about the environments of these boys, and certainly nothing in their environments obviously contributes to their unusual preoccupation. This smells innate to me. (I do not claim to be making a strong case here.)

Anecdotally, I have heard several accounts of first-degree relatives (brothers, or fathers and sons) who discovered that both were crossdressers. The discovery was invariably after both relatives had a great deal of cross-dressing experience that they had hidden from each other. This smells genetic to me. Again, though, this is not meant to be a strong argument.

Autogynephilic cross-dressing usually begins in late childhood or early adolescence, but this does not mean that it is not biological. (Pubic hair also begins at adolescence.) Some autogynephiles claim that they have early memories of their condition, such as the desire to be female. I have tended to be skeptical about these memories, but a recent case seen by psychologist Ken Zucker at the Clarke Institute has made me more open-minded. This was a three-year-old boy whose mother had brought him in to the clinic because of his cross-dressing, which she first observed at around age two. According to the mother, the boy wore her or his sisters' underwear, lingerie, slips, and nighties. The mother also reported that (at age three!) he got erections when looking at women's clothing in magazine advertisements, and he would demand that she buy the items he was viewing. His cross-dressing was sporadic, rather than continuous, and it did not appear to reflect early femininity—he did not say he wanted to be a girl or have other feminine interests, for example. The most fascinating development came when Zucker interviewed the father, who admitted that he had crossdressed erotically since adolescence. There was no indication that the boy had ever seen his father do this or had any opportunity to learn the behavior from him. I predict (as does Zucker) that when he grows up, the boy is very likely to have some variety of autogynephilia. His early onset also smells biological, though as I stressed, early onset is not a necessary component of innate behavior.

Highly relevant to the nature-nurture question is whether autogynephilia has occurred in most cultures and times. In fact, there is only very limited evidence about its occurrence prior to Magnus Hirschfeld's classic work, *Die Transvestiten*, published in 1910. There are a few more-or-less definitive accounts, such as the Abbé de Choisy, who lived in France from 1644 until 1724. Although the historical record does not document Choisy's masturbatory habits (he was, after all, a cleric), it is clear that he was a heterosexual cross-dresser. He was romantically drawn to women, whom he preferred dressed as men. In fact, he once arranged a marriage ceremony in which he dressed as the bride, and the woman as the groom. He clearly experienced cross-dressing, and particularly being admired as a woman, as erotic. He had periods in which he felt guilty about his unusual preoccupation and purged, just as contemporary cross-dressers do.

The cross-cultural occurrence of autogynephilia has not been well established (in contrast to homosexual transsexualism, which has been). This is not surprising. It is probably rare, secretive, and poorly understood. On the other hand, I expect that it occurs everywhere. Blanchard has seen autogynephilic transsexualism in immigrants from Europe and Asia.

In order to progress scientifically toward the causes of autogynephilia, it will be useful to keep in mind that autogynephilia seems to be a type of *paraphilia*. Paraphilias comprise a set of unusual sexual preferences that include autogynephilia, masochism, sadism, exhibitionism (i.e., exposing one's genitals to strangers for sexual excitement), frotteurism (rubbing oneself against strangers, such as in a crowded bus, for sexual excitement), necrophilia, bestiality, and pedophilia. Because some of these preferences (especially pedophilia) are harmful, I hesitated to link them to autogynephilia, which is not harmful. But there are two reasons to think that these sexual preferences have some causes in common. First, all paraphilias occur exclusively (or nearly exclusively) in men. Second, paraphilias tend to go together. If a man has one paraphilia, then his chances of having any other paraphilia seem to be highly elevated. The best established link is between autogynephilia and masochism. There is a dangerous masochistic practice called "autoerotic asphyxia," in which a man strangles himself, usually by hanging, for sexual reasons. Although autoerotic asphyxiasts arrange an escape hatch—for example, a well-placed stool they can stand on before it's too late—sometimes things go wrong. Perhaps 100 American men per year die in this way. About one-fourth of the time, these men are found wearing some article of women's clothing, such as panties. There is no obvious reason why autoerotic asphyxia should require cross-dressing. Apparently, these men are both masochistic and autogynephilic. Cross-dressing has also been linked to sexual sadism—although most autogynephiles are not sexual sadists, they are more likely to be sadists compared with men who are not autogynephilic.

Paraphilias tend to seem bizarre to typical gay and straight people, whose sexual desires are primarily directed toward conventional sex acts with adults. Social explanations of paraphilias tend to be strained and unconvincing. What kind of experiences would make men risk their lives to become sexually aroused from being strangled while wearing panties? I'm betting on biology.

My gut feelings may say as much about my biases as they do about the evidence, which is admittedly scanty. Other people might look at the same evidence and reach the opposite conclusion. However, no one could honestly and competently say that we are anywhere close to understanding the causes of autogynephilia, or more generally, paraphilias.

★★★★★★★★

"Most gender patients lie," says Maxine Petersen, the ace gender clinician at the Clarke. One common lie among autogynephiles, according to Petersen, is that they are homosexual rather than heterosexual. The motivation for that lie is probably the fear that a gender clinic will deny them a sex change if they are determined to be heterosexual. And indeed, some psychiatrists have taken the position that nonhomosexual transsexuals are uniquely inappropriate for sex reassignment because they are not "true" transsexuals.* Autogynephiles who claimed to be homosexual transsexuals could account for the apparent cases of homosexual transsexuals who practiced erotic cross-dressing. Other common lies, according to Petersen and others, include an exaggeration of early femininity. This might in some cases have the same motivation.

The most common way that autogynephiles mislead others is by denying the erotic component of their gender bending. For example, when Stephanie Braverman lectures to my human sexuality class, she does not even mention her history of masturbating while cross-dressed. When I spoke at a meeting of Chicago cross-dressers, the men became clearly uncomfortable when I brought up the erotic component of their activity, preferring instead to attribute it to their inner femininity. When I pointed this out, one cross-dresser said "I wear feminine clothing because I feel feminine, and I can't help getting aroused because the clothes are sexy. Any man would." I don't think so. But you can judge. Here is one of the passages that aroused the cross-dressers in Blanchard's study. See if you think it is sexy.

> You have plenty of time to dress this evening. You slip your panties over your ankles and pull them up to your waist. Sitting on the edge of your bed, you put on a pair of sheer nylon stockings. You fasten the stockings with the snaps of your lacy garter belt. You slip your arms through the straps of your brassiere and reach behind you to fasten it. You put on your eye shadow,

---

*The Clarke Institute does not discriminate against autogynephiles and, indeed, Blanchard wrote an uncharacteristically impassioned passage in one article urging readers not to use his findings to justify such discrimination. However, as recently as 1989, Blanchard and his colleagues from the Clarke opined that "heterosexual applicants for sex reassignment should be evaluated with particular caution" because of an increased likelihood of postoperative regret.

mascara, and lipstick. Lying on your bed, you look up at your reflection in the large mirror on the ceiling.

Why do some autogynephiles deny the sexual component of their condition? One reason, again, is the real or imagined treatment implications. Some psychiatrists refuse to recommend for sex reassignment any man who has had even one incident of erotic cross-dressing. But this fear surely cannot explain the resistance of Stephanie Braverman and the cross-dressers at the meeting—they are not trying to become women.

Perhaps the major reason is shame and assumed social reaction. The physician Harry Benjamin, who popularized the word "transsexual," noticed early on that cross-dressers, and especially cross-dressers in organizations trying to influence the public, tend to deemphasize the erotic element. He suggested that they do this in order to be more accepted by others. Today, public statements by those who call themselves "transgendered" (who are almost all autogynephiles rather than homosexual transsexuals) rarely acknowledge any erotic component of "transgenderism."

There is also a more personal motivation to deny the erotic component of autogynephilia. Anne Lawrence put it this way:

> I imagine most men would be humiliated to admit that dressing in women's clothing is a sexual kick, and even more humiliated to admit that doing so, or fantasizing doing so, is obligatory for climax some or all of the time. Just dressing in women's clothing is shameful enough; but having one's sexual potency contingent upon such an unmanly, "ridiculous" crutch would be almost impossible to admit. Moreover, for anyone who thinks about it, the whole experience of reliance on paraphilic behavior or fantasy for arousal is rather tragic and lonely: it cuts one off from intimate contact during partnered sex, because one is (at least mentally) often making love to oneself rather than to one's partner. Better not to admit this to anyone—especially to one's wife. I think that if the wives of heterosexual cross-dressers knew what their husbands were really thinking about at the moment of climax, they would be appalled. (Of course, this might apply to the wives of other straight men as well; but it's one thing to learn he's fantasizing about making love to Claudia Schiffer, and another to learn he's fantasizing about being forced to wear a French maid's outfit.) On the other hand, to at-

tribute one's cross-dressing to a desire to express one's "feminine side" is much more acceptable. Though the behavior may still appear ridiculous, the putative rationale allows the cross-dresser to portray himself as multifaceted, courageous, and even empathic with his spouse. That's a far easier script for most men to follow.

In my experience, most laypeople are happy to accept the "I'm a woman in a man's body" narrative, and don't really want to know about autogynephilia—even though the preferred narrative is misleading and it is impossible to understand nonhomosexual transsexualism without autogynephilia. When I have tried to educate journalists who have called me as an expert on transsexualism, they have reacted uncomfortably. One said: "We just can't put that into a family newspaper." Perhaps not, but then they can't print the truth.

There is one more reason why many autogynephiles provide misleading information about themselves that is different than outright lying. It has to do with obsession. Something about autogynephilia creates a need not only to enact a feminine self, but also to actually believe in her. It seems important to them to emphasize the permanence of the feminine self as well as her primacy: "I was always feminine, I just managed to hide it. I became a Green Beret as a defensive response to my femininity." In such accounts, the feminine self is the real self; the masculine self is the creation. (I have been arguing that the opposite is closer to the truth.) *Intersexuality* refers to congenital conditions in which biological sex is ambiguous, usually due to hormonal or genetic problems. Cheryl Chase, the intersex activist, told me that transsexuals frequently join intersex groups because they are convinced that they are also intersexual. In most cases, they are not. I assume that these are autogynephilic transsexuals who want to believe that there is a real biological woman inside them as well as a real psychological woman.

The self-presentational deceptiveness of some autogynephiles is a main reason why autogynephilia was not understood until recently. Many clinicians—even some who write books—have taken the information that transsexuals tell them at face value. I recently attended a talk by a well-known psychologist at an academic sex conference in which she presented a case that was clearly autogynephilic (he'd been married and was in his late 40s, among other signs). However, she spoke not one word about her patient's sexual fantasies, dwelling instead on the usual "woman trapped in a man's body" story. Blanchard's ideas have not yet received the widespread attention they deserve, in large part because sex researchers are not as scholarly as they should be and so don't read the current scientific journals.

And although Blanchard's ideas are fundamental to an understanding of transsexualism, they might not matter that much for helping transsexuals, which most clinicians have as their first priority. With luck, the next revision of the *DSM* will distinguish "homosexual" from "autogynephilic" transsexualism. But will popular features on "the transgendered" begin to mention the teenage masturbatory cross-dressing? Will "The Cher Mondavi Story" become a made-for-television movie co-starring "Robot Man?" Probably not, and it is a pity. True acceptance of the transgendered requires that we truly understand who they are.

CHAPTER 10

In Search of Womanhood and Men

I wrote first about autogynephilic transsexualism because it is so much less familiar, and harder to grasp, than homosexual transsexualism. As I mentioned, most people have homosexual transsexuals in mind when they think of transsexuals at all. But homosexual transsexualism is also an interesting, complex condition, and as is the case with autogynephiles, there are misconceptions about homosexual transsexuals. My own recent research has focused on the homosexual type. Oddly enough, most of the homosexual transsexuals I have met, I met through Cher, who is the other type of transsexual. This is odd because, as I have mentioned, in the United States and Canada these days, homosexual male-to-female and autogynephilic transsexuals do not run in the same circles. None of the homosexual transsexuals I asked had ever met a transsexual like Cher before. (Homosexual transsexuals do socialize with each other. Most of "Cher's girls" had met each other before meeting Cher.)

Cher's connection with the homosexual transsexuals is her doing. Early in her transition, after she was diagnosed as "transsexual," she decided that she wanted to try the "major leagues." (Cher's frequent allusions to sports and automobiles are sufficient to indicate that she is not a homosexual transsexual.) So she went to gay clubs and gay bars looking for accomplished transsexuals—those who looked and acted like attractive, sexy women. She watched them, befriended them, and learned from them. When I met her, she not only lived with a homosexual transsexual, but her best friend was one, and she was advisor, confidante, or chauffeur to several others. Her friends' experiences, contrasted with her own, have made her an astute observer of their differences. When I asked her opinion about the main difference between transsexuals "from gay versus straight backgrounds" (as she puts it), she said, "Gay transsexuals are boy crazy."

One implication of Cher's assertion is that homosexual transsexuals are like gay men. Many of the facts discussed in the last section on gay men apply to homosexual male-to-female transsexuals. For example, the causes of homosexual transsexualism are largely the causes of homosexuality. To be sure, only a small minority of gay men become transsexual, but homosexual transsexuals are a type of gay man. Richard Green began his important study of feminine boys (discussed in the last section) precisely to see if he could predict which boys would become transsexual adults. Sensibly, after hearing the memories of transsexual patients, he sought extremely feminine boys. In adulthood, most of these boys were gay men, and only one of the sixty in his study was clearly transsexual. Evidently, something prevents most very feminine boys from becoming transsexual. In order for a feminine boy to become transsexual, something extra must happen.

What is the "something extra"? Ken Zucker, whom we met in Chapter 2, has tried to predict which boys with gender identity disorder (GID) would still have the disorder when they become adolescents. Adolescents with GID are much rarer and presumably much closer to being transsexual. Zucker found several predictors of adolescent GID: lower IQ, lower social class, immigrant status, non-intact family, and childhood behavior problems unrelated to gender identity disorder. Obviously, none of these factors can be considered very specific. Parental divorce and low social class are both very common, and most males who experience them do not become transsexual. The factors do, however, suggest a common theme: early adversity. I will speculate later about what this might mean.

When I have discussed the theory that homosexual transsexuals are a type of gay man, I have met resistance. I was surprised at this, for the idea is neither new nor, it seemed to me, controversial. Some of the resistance was emotional. People who believe that homosexuality is not a disorder tend to dislike the implication that a subset of homosexuals are disordered. I think that this is a bad reason to object to the theory, no better than to object to the theory that autogynephilia is a form of heterosexuality because autogynephilia can be considered a disorder.

Another reason why people have difficulty with the notion that homosexual transsexualism is a form of homosexuality is that at their endpoints, the two conditions seem quite different. The picture of the muscular gay man in leather looks quite different from that of the shapely postoperative transsexual in an evening dress. But look at the entire trajectory. As I have emphasized, as boys, some gay men look just like transsexuals. Nearly all homosexual transsexuals go through a stage in which they are "gay boys," feminine to be sure, but not distinctly more feminine than many gay boys who will become gay men. And there is that "missing link" between transsexuals and ordinary gay men—the drag queen. Drag queens are gay men who cross-dress occasionally but who have no intention of changing their sex, and who do not take measures to physically feminize their bodies. Unlike heterosexual cross-dressers, drag queens do not become sexually aroused by dressing in women's clothes. Some drag queens are transsexuals who have not yet accepted it, but for others, occasional cross-dressing is as close to female as they will ever get. In our study, we found that drag queens ranked between gay men and transsexuals on a number of traits related to femininity. Also, they liked their penises much more than transsexuals did.

The other reason some people object to linking homosexual transsexualism with homosexuality is, they argue, that this confuses sexuality with gender. The standard transsexual narrative says that transsexualism is not about sex but about "gender identity," or the internal sense that one is a man or a woman. According to this narrative, transsexuals want to change their sex because their sense of self disagrees with their bodies, not because they have any unusual sexual preferences that depend on a sex change. While the first part of this explanation sometimes may be true, the latter is not. It should be clear by now that the "gender, not sex" part of the transsexual narrative is false for autogynephiles, whether they are transsexuals or merely cross-dressers. Autogynephilia is a very unusual sexual orientation (towards oneself as a woman), and it is usually accompanied by specific and intense sexual imagery. But it would be a mistake to think of autogynephilic transsexualism as the sexual type of transsexualism, and homosexual transsexualism as the type that is solely a disorder of gender identity. Homosexual transsexuals are in their own way just as sexually motivated as autogynephiles.

★★★★★★★★

There is no way to say this as sensitively as I would prefer, so I will just go ahead. Most homosexual transsexuals are much better looking than most autogynephilic transsexuals. There is the rare exception, but for the most part, autogynephilic transsexuals aspire (with some success) to be presentable, while homosexual transsexuals aspire (with equivalent success) to be objects of desire. Homosexual transsexuals have been models and actresses. For example, the model, Tula, was in several movies and posed for *Playboy* before she was exposed as a transsexual.

There are three reasons why homosexual transsexuals are better looking. First, they are typically younger when they start transitioning. This almost certainly helps prevent some of the masculinization that might have occurred had they waited 8 to 10 years, when they would be the same age as the typical autogynephile. Second, they want to attract men, and they get constant feedback (in the form of propositions from men and mostly unsolicited critical advice from their transsexual sisters) about how they are doing. This allows them to hone their presentations faster than the autogynephilic transsexual, who has spent most of her femme life looking at a mirror by herself.

Finally, homosexual transsexuals are better looking because homosexual men who want to be women tend not to enact that desire unless they can pull it off. The standard transsexual story implies that the transsexual is so dissatisfied with her incorrect male body that she cannot wait to discard it, regardless of how good she will look as a woman. This is another place where the standard narrative is wrong, at least about homosexual transsexuals. I have begun asking the homosexual transsexuals I meet whether, if they had looked awful as women, they would have transitioned to full-time females. Most have said "No," and no one has answered with an unambiguous "Yes." Extremely muscular and masculine looking homosexual transsexuals probably choose not to transition, but instead remain among gay men, who value their masculine looks. Blanchard has found that homosexual transsexuals tend to be physically smaller than their autogynephilic sisters, which is consistent with just this sort of self-selection. The "Before" and "After" pictures they have shown me also support my thesis. As men, the homosexual transsexuals look and act extremely feminine, and that presentation is not very marketable among gay men. They are far more fetching as women.

Kim, whom I mentioned at the start of this section, exemplifies the dilemma that some homosexual transsexuals face. Recall that when I first saw Kim, she was at Crobar with a very handsome and muscular man, and I thought they looked sufficiently like a beautiful heterosexual couple that I refrained from approaching her. Sure enough, though, when I told my transsexual informants about her later, they recognized the description and claimed Kim as one of their own. I arranged to interview her for the study we were conducting. When she came to my laboratory, my initial impression was reconfirmed. She was stunning. (Afterwards, my avowedly heterosexual male research assistant told me that he would gladly have had sex with her, even knowing that Kim still possessed a penis.) Yet despite her appearance, Kim was the most ambivalent homosexual transsexual I had met. She didn't know if she wanted the operation. It turned out that the handsome man with her at Crobar is a gay man, who enjoys her company and being seen with her, but who of course could never be attracted to Kim the way she looked that night. Kim had fallen in love with him, however. Because she knew that she could never have him as a transsexual or a woman, she was considering reversing her transition and becoming a man again. In her case, this would have meant removing breast implants and silicon in her hips. She also said she would probably have to hit the gym to bulk up. As she told me of her dilemma, I increasingly wondered what she could be thinking. I could not believe that Kim could ever be attractive enough as a man to attract the likes of the gay man I saw. Such a sexy woman could not possibly make the kind of masculine, muscular man that gay men tend to prefer. I silently predicted that Kim would come to her senses, let her man go, and embrace the femme fatale she was well on her way to becoming. So far, I am half right. Kim is no longer seen with him, and she is still a nascent woman, but she has not yet gotten surgery.

Kim's story shows that sex reassignment is not necessarily an inevitable, unwavering goal for the homosexual transsexual. Rather, sex reassignment has a rational choice component: "Can I make it? Will I be happier as a female? Will I be more successful getting straight men as a woman than I am at getting gay men as a man?" (The last decision has to be weighted by a particular transsexual's degree of preference for straight versus gay men. Most vastly prefer straight men.) This could help explain why the large majority of boys who start out looking transsexual ultimately do not pursue sex reassignment. Some doubt they can be attractive women.

This is a significant difference between the homosexual and autogynephilic types. In making the decision whether to undergo sex reassignment, the autogynephiles do not seem to dwell much on whether they can attract mates. Even autogynephiles who worry that they cannot pass as women are concerned more about stares at the grocery store than about a lack of stares at the cocktail lounge. This makes sense. The autogynephile's main romantic target is herself. This is also consistent with my intuition that autogynephilia is a very internally driven condition, much less susceptible to the kind of rational analysis that homosexual transsexuals seem to engage in.

****

Alma is a 40-year-old Latina homosexual transsexual who got her sex change in her mid-30s—quite late for the homosexual type. She looks great, and works as a call girl out of her condo, which she owns. Alma has seen more than one era of transsexuals. Her brother (whom she calls her "sister"—there is no avoiding this kind of confusion) was also a transsexual, who saw the heyday of New York's Studio 54 before dying of AIDS in the early 1980s. Alma has seen many a transsexual come and go, and the first thing that she thinks of that most have in common is that they are outcasts. They are outcasts as children because of their extreme femininity. They mostly come from poor, broken families, and family rejection is common. The gay community rejects transsexuals, according to Alma, because "they're jealous that we get to have sex with straight men."

Alma has also noticed, as I have, the large number of Latina transsexuals. In Chicago, there are several bars that cater to Latina transsexuals. About 60 percent of the homosexual transsexuals and drag queens we studied were Latina or black. The proportion of nonwhite subjects in our studies of ordinary gay men is typically only about 20 percent. Alma says she thinks that Hispanic people might have more transsexual genes than other ethnic groups do. Another transsexual, remarking on the same phenomenon, attributed it to ethnic gender roles: "My culture is very macho and intolerant of female behavior in men. It is easier just to become a woman."

I am not sure about the validity of all of Alma's observations, much less her theories, but there is clearly something to the idea that homosexual transsexuals are used to living on the margins of society. They have, in fact, had to learn to cope with rejection and disapproval since childhood, because of their extreme femininity. And they have not had the advantages that tend to instill respect in the social order. The early chaotic backgrounds of so many homosexual transsexuals might help explain why they do not defeminize the way that most very feminine boys do. A feminine boy from a middle-class or upper-middle-class family (such as Danny's) has more motivation to "hang in there" until he normalizes his gender role behavior, because he has a good chance at a conventionally successful future. Defeminization might also require more ambition and family support than some homosexual transsexuals possess.

Most homosexual transsexuals have also learned how to live on the streets. At one time or another many of them have resorted to shoplifting or prostitution or both. This reflects their willingness to forgo conventional routes, especially those that cost extra time or money. Homosexual transsexuals tend to have a short time horizon, with certain pleasure in the present being worth great risks for the future.

Prostitution is the single most common occupation that homosexual transsexuals in our study admitted to. About half of them have worked as prostitutes at some point. In Chicago, the entry-level position is as a female-impersonating streetwalker who works the area of Broadway that is mostly gay after dark. (Their customers, of course, are not gay men. They are either unwary straight men or men looking for she-males.) This kind of prostitution is dangerous, especially for transsexuals, whose customers sometimes do not know what they are. They often form relationships with street hustlers or ex-cons. The rate of HIV infection among transsexual streetwalkers is very high, partly due to the high rate of intravenous drug use.

The more resourceful and attractive transsexual prostitutes are call girls. Before their sex reassignment surgery, they advertise as transsexuals. There is, in fact, a market for the services of preoperative transsexual prostitutes, and I will discuss this later. After surgery, many transsexual call girls continue in the business. Alma's friend, Juanita, is a very attractive postoperative transsexual who has worked as a call girl both before and since her operation. Juanita differs from genetic female prostitutes because she asks men to describe themselves on the phone before she makes an appointment with them. In doing so, she is trying to determine whether their appearance will be acceptable to her. For example, she rejects obese men. She also admits that she finds some of the men who patronize her attractive, and enjoys sex with them. She doesn't tell them, though, because she doesn't want them to try to get sex for free. Although Juanita says she would like to switch occupations, she does not feel degraded and guilty about what she does for a living. I suspect that this reflects an aspect of her psychology that has remained male. When we ask transsexuals about their level of interest in casual sex, they respond pretty much like gay men and straight men, all of whom are more interested than either lesbians or straight women, on average. Although Juanita is so feminine in some respects, even some behavioral respects, her ability to enjoy emotionally meaningless sex appears male-typical. In this sense, homosexual transsexuals might be especially well suited to prostitution.

As for shoplifting, homosexual transsexuals are not especially well suited as much as especially motivated. For many, their taste in clothing is much more expensive than their income allows. Transsexual call girls are among the few who can afford expensive clothes. In female impersonator shows, transsexuals often wear designer gowns, which are widely believed (by other transsexuals) to have been acquired via the five-fingered discount.

Living on the edge is more out of necessity than desire. Most of the homosexual transsexuals I talked to had similar dreams for the future. They wanted to get their surgery (if they had not yet had it) and meet a nice, attractive, and financially stable heterosexual man who would marry and take care of them. This is obviously similar to the hopes of many non-transsexual women. When I was conducting my study of homosexual transsexuals, I routinely asked them if they knew anyone who had realized this dream. No one did.

★★★★★★★★★

The Baton is Chicago's premier female impersonator club, featuring several past Miss Continentals, including the gorgeous Mimi Marks. The performances consist of lip-synching and dancing to well-known songs, and the intended effect is to awe the audience with the beauty and realism of the female impersonators, who all appear to be women. Mission accomplished. Even the less attractive performers are not so because they look like men, but rather, because they are overweight, or merely plain. They look like women. All the performers I met there label themselves transsexual, and they all love men. They also all still have their penises. Once they have their sex reassignment surgery, they become women, and women cannot impersonate women.

My first time at the Baton, I too was wowed by the accomplished female impersonations. But the most interesting part of the experience involved the audience. One man who sat close to the stage, by himself, was the object of derision by the transsexual MC and several of the other performers, during their acts. The performers made gestures indicating that he was dirty or perverse, while the man gazed up at them, seemingly unfazed. At one break, I overheard one of the performers telling him, exasperated, "Of course I still have it!" Only later, when I spoke to several homosexual transsexuals about it, was I able to surmise what was going on. None of the transsexuals I asked had difficulty interpreting the interaction. It was evidently in the realm of experience of all of them, in one way or another.

The man was at the Baton because he was especially attracted to she-males, or transsexuals who live as women but still have their penises. She-males are most often depicted as mostly feminine individuals, with women's faces, breasts, and absence of facial and body hair, but with functioning and erect penises. She-males are not just an acceptable substitute to this man. They are his preferred targets. Evidently, there is a significant market for she-male sex. Advertisements in pornographic magazines often sell videos or other magazines featuring she-males. About half of the homosexual transsexuals I have met have worked as prostitutes, and the majority of these worked preoperatively as she-males. One study found that among prostitutes' solicitations in a Toronto alternative newspaper, about one in twenty was placed by a preoperative transsexual prostitute.

Who are the customers? Are they gay, straight, or bisexual? Are they merely men looking for something exotic? Ray Blanchard is the only researcher who has studied men who are sexually attracted to she-males. (Blanchard calls men with sexual interest in she-males "gynandromorphophiles." Cher calls them "transie sniffers." I will stick with "men with sexual interest in she-males.") In a content analysis of sexual personals advertisements, he found that about half of men who sought she-males were cross-dressers. Blanchard thinks that a significant number of men who want she-males are "partial autogynephiles"—they are primarily aroused to the image of themselves as she-male. Blanchard says that the men are not gay but are more like "scrambled up heterosexual men." The transsexuals I know who worked as she-male prostitutes confirmed this. "There was nothing gay about those men," said one, who knows plenty about gay men.

There is a rather uneasy symbiosis between the homosexual she-males, on their way to sex reassignment, and the men who want them at that stage. Juanita, who has been a successful prostitute before and after sex reassignment surgery, says simply "You would have to be crazy to prefer being a she-male prostitute." According to Juanita, there were several problems with customers who call on she-males. Most annoyingly, they frequently don't show up for the appointments they scheduled. She thinks the no-shows want something exotic but simply lose their nerve and decide they can't go through with it. Another thing that irritated Juanita about the customers who called on her when she was a she-male was the way they viewed her. "They considered me their little sex toy and assumed that just because I was a transsexual I would do anything kinky. They didn't care about me, or even what I looked like. They just wanted to know if my thing worked." Juanita says that the most frequent unwelcome requests were that she would penetrate them anally, act like a dominatrix, or allow them to cross-dress with her. The most frequent activity that she granted was oral sex (the men sucked Juanita's penis).

Juanita has had her sex reassignment surgery, and now works as a call girl for men who want real women. She does not tell them that she used to be a transsexual. None of her frequent customers from before was interested in continuing with her, post-surgery. The new men are more "intimate," according to Juanita, because they see her as a real person rather than merely a "sex toy." Juanita is a very attractive transsexual, and had the luxury of continuing to work successfully even after she was no longer exotic. Although I have not met one, some she-male prostitutes allegedly delay sex reassignment surgery because they are concerned that their incomes will suffer after they no longer have their penises.

★★★★★★★★

The voice on my answering machine sounded serious, even worried, and I wondered why. Maria had seemed ready and eager for her surgery. She was one of the few homosexual transsexuals I had met who had a conventional job; she was cheerful and not at all ambivalent about the surgery. She knew what she wanted. When I called her back, she asked to meet at a restaurant in Lincoln Park, Chicago's trendy urban neighborhood. When I arrived, she greeted me at the door, and I barely recognized her. She had been quite passable before, but not especially attractive. Now she was the kind of woman that men gawk at (and later when we left the restaurant, they did). She now had very large breasts and an hourglass figure. Her face, which showed light stubble before, was radiantly feminine now. And she had already had one of the best voices I had ever heard on a transsexual. Yet she began by saying she had problems.

Wherever she went, Maria was constantly feeling that people were whispering about her, identifying her as a transsexual. I was quite certain that people were whispering about her, but equally certain that they were not "clocking" her (detecting her status as a transsexual). Then she revealed her current personal situation, which helped explain her paranoia. For over a year she had had a steady boyfriend who did not know that she is a transsexual. She had made up a past life in response to his queries. Her gay brother collaborated with her to convince her boyfriend of the truth of her false past and to hide the true past. She was extremely concerned that her boyfriend would find out, and the constant worry caused tension in their relationship. For example, she was jealous that he would seek a "real woman," although in fact he believed he was already with a "real woman," and they had been fighting.

Maria had met her boyfriend shortly after getting breast implants but before her vaginoplasty. Evidently, many men had made advances at that time, and she chose him because he was good looking and ambitious. She was able to postpone intercourse with him for a few months, meanwhile frantically managing to get her surgery scheduled sooner. She had sex with him sooner than she was supposed to, but had not had any physical problems as a result. Her new body worked well.

She and her boyfriend had talked fairly specifically about a future, including marriage and children. Although aware that she cannot have children, she was willing to adopt. The boyfriend had integrated her into his circle of friends and introduced her to his family, who loved her. Maria had absolutely no intention of telling her boyfriend, ever.

She had cut off, or at least drastically reduced, interactions with her old transsexual friends to reduce the chance of discovery, but had a couple of close calls in public with her boyfriend, once with Cher. Some of her old friends understood and wished her well; others did not. In fact, Maria worried that a resentful transsexual might track down her boyfriend and tell him merely to spoil things for her.

As we spoke, I sympathized with Maria, but I also pondered her and her boyfriend's predicaments. Maria's is clear enough. Ray Blanchard once presented the following dilemma to a prominent and open-minded heterosexual male scientist. Suppose that you met the perfect woman—attractive, sexy, and interested in you—with one catch: She is transsexual. Would you be her partner? The scientist sheepishly admitted that he would not. When I asked Juanita, the sexy transsexual prostitute (now post-op, and not generally open about her past) about the best, and worst, reactions she had had from lovers after she revealed that she used to be a man, she replied "I have really never had a good experience. The men always leave." Juanita's most recent boyfriend confronted her after penetrating her for the first time. Her vagina is shallow, and he concluded that she is not a normal woman. He asked if she is transsexual, and she did not deny it. He ran from her apartment and called her later to say that he could not deal with her revelation just now. She has not heard back from him. Juanita knows only one transsexual who has been with a man for more than a year, and that transsexual's boyfriend pimps for her.

All the homosexual transsexuals I have talked to say that they wish they could find a man they could tell and who would love them anyway, but they all worry that such a man does not exist. And they are all deeply suspicious of men who prefer transsexuals to real women. (These men have something similar to "sexual interest in she-males," and transsexuals find them weird.) There is little incentive for the postoperative homosexual transsexual to be honest.

Cher has made it clear to her friends, such as Juanita, that she disapproves of such deception, and that she intends to be honest with her own prospects. (Cher currently considers herself bisexual, but thinks that she is most likely to become sexually involved with men.) Press her, though, and she will admit that virtue is somewhat easier for her than for her homosexual transsexual friends. Cher is sufficiently "clockable" that she cannot risk not telling. Before she says anything, most men know, or at least suspect. Furthermore, I do not believe that Cher's attraction to men is as intense or as unambiguous as that of homosexual transsexuals. She is autogynephilic, and men's place in her sexual world is complicated. So the loss of a potential sex partner is less of a loss, overall, to Cher than it is to the homosexual transsexuals, who simply lust after men.

I put myself in Maria's boyfriend's place and ask myself if I would want to know. The answer is less clear the longer I contemplate. On the one hand, any person to whom it mattered would seem to have the right to know. On the other hand, this is a man who by all accounts is in love with Maria, and who derives a great deal of satisfaction from being with her. (My impression is that his friends and family believe he is very lucky to be with her.) Even though the couple (if it remains a couple) is destined to be biologically childless, this is less of a problem for most men than for most women, who often become depressed when a couple is infertile. By the kind of utilitarian analysis I am partial to, let us ask which ending would leave the world a happier place: the boyfriend finds out, or he doesn't find out. Assuming that the couple is destined to break up for other reasons (after all, they are only in their early 20s), then surely it is better for both if he does not find out. If they are compatible enough to make a life together, then it is still not clear that he should know. After all, she could eventually reveal that she is sterile without saying why. If having biological children were so important to him, he could end the relationship with only that knowledge.

Maria asked me to talk to her and her boyfriend, to do couples counseling, pretending that I have known her only as a woman. This would serve both the goals of helping their relationship and covering her story. I considered it briefly, then refused. But it was not an easy decision.

Autogynephilic and Homosexual Transsexuals: How To Tell Them Apart

Once you have learned about auto-gynephilic and homosexual trans-sexuals and seen several of each, distinguishing them is easy. If Blanchard and I saw the same 100 transsexuals, I would be surprised if we disagreed on more than 2. But most readers will not have met a single transsexual of either type, and even most clinicians who see gender patients are not used to thinking about them this way. In any case, you cannot simply ask someone "Which type are you?"

I have devised a set of rules that should work even for the novice (though admittedly, I have not tested them). Start at zero. Ask each question, and if the answer is "Yes," add the number (+1 or –1) next to the question. If the sum gets to +3, stop; the transsexual you're talking to is autogynephilic. If the sum gets to –3, she is homosexual.

+1 Have you been married to a woman?

+1 As a child, did people think you were about as masculine as other boys?

+1 Are you nearly as attracted to women as to men? Or more attracted to women? Or equally uninterested in both? (Add 1 if "Yes" to any of these.)

+1 Were you over the age of 40 when you began to live full time as a woman?

+1 Have you worn women's clothing in private and, during at least three of those times, become so sexually aroused that you masturbated?

+1 Have you ever been in the military or worked as a policeman or truck driver, or been a computer programmer, businessman, lawyer, scientist, engineer, or physician?

-1 Is your ideal partner a straight man?

-1 As a child, did people think you were an unusually feminine boy?

-1 Does this describe you? "I find the idea of having sex with men very sexually exciting, but the idea of having sex with women is not at all appealing."

-1 Were you under the age of 25 when you began to live full time as a woman?

-1 Do you like to look at pictures of really muscular men with their shirts off?

-1 Have you worked as a hairstylist, beautician, female impersonator, lingerie model, or prostitute?

Finally, if the person has been on hormones for at least six months, ask yourself this question:

If you didn't already know that this person was a transsexual, would you still have suspected that she was not a natural-born woman?

+1 if your answer is "Yes" (if you *would* have suspected)
-1 if your answer is "No"

Keep in mind that people don't always tell the truth. This inter-view could be invalid if the transsexual is actually autogynephilic but is either (a) worried that you will think badly of her or deny her a sex change if you know the truth, or (b) obsessed with being a "real" woman.

CHAPTER 11

Becoming a Woman

The medical transitioning of transsexuals, from men to women (and the other way, too), is no longer just a curiosity, but a business. If not a big business, it is at least a lucrative business for a few surgeons, who devote their entire practices to it. Some of them have their own Internet websites and distribute videos that describe their services and show their results. It seems to be a rapidly advancing specialty, as well. Cher, who had her genital sex change surgery only eight years ago, notes somewhat enviously that neo-vaginas now look so much more realistic, complete with realistic-looking (and sensitive) clitorises, and labia. Advances are surely driven in large part by the free flow of information. Few brain surgery patients study their options more closely than do transsexuals, who trade not only opinions but also stories and pictures, both informally and on websites. (Next to a close-up photograph of a neo-vagina spread by the patient's fingers: "This is a fairly typical Dr. M. result, with a well-defined clitoris and nice thin labia. The urethral opening here is a little lower and harder to see than in some examples: Dr. M. seems to be trimming his urethras shorter recently.") The high-tech websites are nearly all maintained by autogynephilic transsexuals, but homosexual transsexuals spread the word about the same surgeons, and the surgeon's office is one place where homosexual and autogynephilic transsexuals might well meet. Here are the main medical procedures that male-to-female transsexuals undergo, in rough chronological order in which they are typically undertaken, with rough costs.

Start with electrolysis, to get rid of the beard. (Electrolysis of body hair is a lower priority, because it recedes some with hormonal treatment and in any case can be hidden.) Autogynephiles prefer to do this while still in the male role. Homosexual transsexuals, because they are younger and possibly have less facial hair to begin with, tend to switch roles first. Weekly time can range between one to more than five hours at $40 to $100 per hour. Completion may require less than 100 to more than 700 hours, with an average between 200 and 300 hours. Typical total electrolysis costs range between $4,000 and $16,000. Recently, some surgeons have recommended getting electrolysis on the scrotum between the legs, as well, because this skin is often used to line the neo-vagina, and should be hair-free. Electrolysis hurts and leaves red blotches on the skin for a while after each session.

Next, hormones. Female hormones (synthetic or "natural" estrogens, the latter taken from animal urine) are taken either orally, by transdermal skin patch, or by injection, for the rest of the transsexual's life—assuming that she goes all the way. Also, while the transsexual still has her testes, she usually must take some kind of anti-androgen hormone as well; this can be discontinued after she is castrated (which usually happens during sex reassignment surgery). Hormone therapy is not typically very expensive—less than a couple of dollars per day— and can be had either through a physician or without a prescription in Mexico or even by mail order via the Internet.

If given early enough, hormones prevent masculinization of facial and body hair, and facial and body skeletal structures. Early enough for complete prevention is prior to puberty, and this does not happen in this country. (In the very liberal Netherlands, hormone therapy to delay puberty sometimes is given in early adolescence.) But even in the late teens and early twenties, hormone therapy can prevent a significant amount of masculinization that would otherwise occur. This is one reason why homosexual transsexuals tend to be more convincing as women compared with autogynephilic transsexuals, who tend to be older before starting hormones.

Hormone therapy causes breast growth that is typically about one or two cup sizes less than sisters and mothers reach. Male sex drive decreases (and this is often experienced as a relief). Fat is redistributed, causing the face to assume a more feminine shape. Fat leaves the waist and moves toward the hips and buttocks. Body hair growth slows, becomes less dense and lighter colored (but not on the head, face, or pubic area). Many transsexuals say that female hormones make them feel better, and less depressed. Some transsexuals say that female hormones make them behave more female-like. Some say that it makes them more attracted to men, for example, and Cher believes that female hormones make her hold a cup like a woman rather than like a man. Some of these psychological "effects" of hormone therapy are probably placebo effects, although it is not unlikely that others are real. The worst potential side effect of hormone therapy is blood clots that can travel to the lungs, where they can be fatal. Luckily, this side effect is rare.

With electrolysis and hormones, the other thing to get started on early is the voice. Female hormones do not feminize the male voice, once it has changed. The voice is a big hindrance to many transsexuals in their quest to pass. It is particularly difficult to pass on the phone, when they cannot convey their otherwise (in many cases) very feminine presentation. The medical solution to the voice problem, "voice surgery," involves tightening of the vocal cords so that the pitch of the voice is elevated. It is convenient to get a tracheal shave at the same time (Adam's apple reduction), for a total of $4,500 or more. However, voice surgery is still not considered very reliable—it has produced too many bad outcomes, such as hoarsenes—and most transsexuals opt for a few sessions with a voice therapist. The voice therapist teaches ways to sound more like a woman. The most important and obvious focus is raising the pitch of the voice to be as high as possible. Singing is good practice for this. Even aside from pitch, men and women talk differently. Women have more jumps in frequency than men, conveying a more singsong effect. They have more precise articulation. They ask more questions and talk about feelings more. In my experience, the transsexual voice remains the most problematic piece of the feminine puzzle. I have met many transsexuals whose physical appearance does not give them away, but I have met only a few whose voice provides no clue.

Get rid of the beard, grow long hair, and put on a dress and even with good breast growth, some transsexuals look like men in dresses. Male and female faces differ, and everyone sees the face. Men, especially older men, have higher hairlines, broader chins, "brow bossing" (a prominence of the male brow ridge), lower eyebrows, narrower cheeks, and more prominent, angular noses. All these masculine features can be somewhat feminized with surgery. Facial plastic surgery is expensive, potentially the most expensive thing that a transsexual will buy. Total costs can exceed $30,000, but this varies greatly. Some transsexuals (especially the homosexual type) need relatively little, and others need a lot of work.

Although hormones cause some breast growth, many transsexuals elect to get breast implants as well. Homosexual transsexuals almost invariably do this, and their tastes run large. They want to be noticed. (One homosexual transsexual I know got her sex-change surgery several months earlier than she had originally planned because after she got her breast implants, she immediately obtained a boyfriend who wanted to have sex. Another told me that the most unrealistic aspect of the portrayal of Dil, the transsexual in *The Crying Game*, was that she had not managed feminine breasts.) This surgery is well known to genetic women these days, and costs about $5,000. One surgeon offers a discount if the implants are done at the same time as genital surgery (not recommended by some, because there is then no comfortable part of the body to put weight on). More than one transsexual told me that the aftermath of breast implant surgery was far more painful than that of genital surgery.

Women's hips and bottoms are wider than men's, so some transsexuals get silicon injections there. Silicon injections can be dangerous. Silicon can enter the bloodstream and travel to the lungs, causing a fatal embolism. Also, because the silicon is loose rather than enclosed in surgical implants, there is concern that the silicon will eventually migrate to other places and look bad. (I have been unable to find anyone to whom this has happened, but it is well documented in the medical literature.) Many people consider physicians who administer silicon injections to be disreputable. Homosexual transsexuals have more motivation to attract men in the short term and seem less concerned with long-term consequences, so they are more apt to get the silicon injections. Ideally, these should be done in series, waiting for each layer to harden before putting another one on. Currently in Chicago, the person who does this procedure for most transsexuals is, herself, a transsexual who works out of her apartment. Facial injections cost $125, hips $600, and bottoms $400.

The most exotic procedure—though not necessarily the most expensive—is vaginoplasty, or the construction of a neo-vagina. There is more than one way to accomplish this. In any method, the first step is to remove the testicles and the erectile tissue (insides) of the penis. A "vaginal" cavity is created between the urethra (urinary tube) and the rectum. In the most common form of the operation, the penile skin is inverted to form the lining of the neo-vagina. If the patient has a short penis (less than 5 inches), the surgeon can graft skin onto the new vagina to lengthen it. (If the scrotum has been cleared of hair by electrolysis, this skin can be used.) The glans (head) of the penis is used to

construct the clitoris. Because the glans contains the nerves that produce most of the penis's erotic sensations, the neo-clitoris is usually
sensitive (just as a genetic woman's is). A part of the scrotum is used to
make the labia (vaginal lips). These days, the best surgeons offer a
second, optional, labiaplasty operation, in which the labia are thinned
and a clitoral hood is formed, making the overall appearance generally
more realistic. The most interesting variation in the procedures I have
outlined is that some surgeons can use a segment of bowel tissue to
line the vagina. According to some reports, this makes the vagina
lubricate naturally, but this kind of vaginoplasty is more expensive and
carries a greater risk of complications. Typical vaginoplasty fees range
from less than $8,000 to more than $15,000, and some surgeons charge
as much as $30,000. The optional labiaplasty is only about $3,000.
After the big operation, the patient must stay in a hospital for about
three days (included in the total cost), and when she can, she must
dilate her new vagina regularly, with dildo-like plastic rods.

I suspect that 10 years from now, this section of the book will have
to be much longer to provide even superficial coverage of available
options. And some of the procedures I have described will seem primi-
tive by comparison. But even in the recent past, desperate transsexuals
without much money have been subject to far more rudimentary and
dangerous surgery by quacks eager to exploit them. One notorious
doctor, John Ronald Brown, was named one of the nation's worst
surgeons by *Vogue* in the late 1970s, served time for illegal medical
activities, then reopened his sex change shop in Tijuana. Transgender
activist Dallas Denny, who wrote an exposé on Brown, described the
"Tijuana experience":

> Many of the transsexual people who went to Mexico for gender reassign-
> ment surgery in the seventies and eighties wound up mutilated, with geni-
> talia looking like they belonged to one of the creatures in the bar scene in
> "Star Wars," and not like something likely to be found on a human being of
> either gender. Some of these people, expecting vaginoplasties, received
> simple penectomies, leaving them looking somewhat like a Barbie doll.
> Others ended up with something that looked like a penis that had been

split and sewn to their groin—which is essentially what had been done. Some ended up with vaginas which were lined with hair-bearing scrotal skin; these vaginas quickly filled up with pubic hair, becoming inflamed and infected. Some ended up with peritonitis, some with permanent colostomies. Some ran out of money and were dumped in back alleys and parking lots to live or die. Some died in those parking lots or back in the States, of complications from the surgery.

In 1998 Brown was arrested for performing an illegal amputation that led to the death of an elderly man. Speculation is that the man had an "amputee fetish" (yes, this exists, and can be explored thoroughly on the Internet) and found Brown after legitimate surgeons refused him.

****

Sex reassignment surgery is the easy part. The difficult part for many transsexuals is the social transition that usually precedes surgery by several years, and that continues for years afterwards. This transition is especially difficult for autogynephilic transsexuals, and is worst for those with wives and children and jobs that require them to interact with the public. Most physicians and mental health professionals who work with transsexuals adhere to the Standards of Care for Gender Identity Disorders, promulgated by the Harry Benjamin International Gender Dysphoria Association (named for the revered father of transsexualism). Among other things, the Standards of Care specify that prior to medical treatment (such as hormones or sex reassignment surgery), transsexuals must participate in "real-life experience," living full time as the sex they will become.

The Clarke Institute of Psychiatry has a very conservative real-life experience requirement. Transsexuals who want to become women must first live for a year as women before receiving hormone therapy. (This is motivated by concern for genetic females who want to become men. Once they receive testosterone, their voices will permanently deepen. Genetic men who receive female hormones do not risk analogous permanent effects. However, Blanchard does not want to risk accusations of gender bias, so he holds both female-to-male and male-to-female transsexuals to the same requirements.) During this time, they must work, volunteer 20 hours a week, or attend school full time, while maintaining a female identity. They must submit proof—either tax forms or letters from bosses or supervisors—that they are known by an unambiguously female name. "Pat" would not count; "Patricia" would. Once they begin hormone therapy, transsexual patients must live for another year (two total) as women before receiving official authorization for sex reassignment surgery. Currently, the Clarke Institute arranges to have successful applicants' sex reassignment surgeries performed in England by surgeons Blanchard thinks are top notch.

Many transsexuals find the Clarke Institute's lengthy real-life experience requirement to be onerous. They believe they should be eligible for hormones and surgery much sooner. There have even been a few cases in which impatient patients mutilated their own genitals. (These have all been autogynephiles, according to Blanchard.) Indeed, there is no hard evidence in favor of the Clarke's policy. (To get hard evidence, one would have to randomly assign transsexuals either to the two years of real-life experience or to a shorter requirement, and then follow them up to see which group fared better. No one has done this.) There is new evidence that transsexuals who have had real-life experiences as short as six months can do fine after surgery. Still, the Clarke gender staff thinks the two year period is a good idea. Blanchard simply believes that the likelihood of regrets is too high with a shorter period. Maxine Petersen emphasizes the importance of learning:"The feeling of belonging to a different gender and the actual experience of what it is like to belong or to live in that gender role and be accepted as female are quite different. Until one has done it the idealized existence is likely to dominate." In part, Petersen is referring to experiences that all women confront, such as being patronized by garage mechanics. She is also referring to transsexual-specific experiences. Many transsexuals will have to contend for the rest of their lives with other peoples' stares, smirks, and whispers, and a real-life experience presents them with the opportunity to know if they can live with that. Most homosexual male-to-female transsexuals do not know about the Standards of Care, much less attempt to adhere to them. Yet even for them, there must be an ultimate decision to stop using a male identity and to adopt a female identity full time. My impression is that this is an easier, less traumatic transition for them than for the autogynephiles. For example, most homosexual transsexuals I talked to felt sufficiently confident about their appearance when they transitioned full time that this was not a source of major discomfort. On the contrary, by the time they go full time, most homosexual transsexuals have had feedback—from other transsexuals and from straight men—that they can pull it off. The two groups most likely to have problems with their transition—family and employers—are less difficult for homosexual transsexuals. Homosexual transsexuals are more likely to be estranged from their families, in which case they care less what their families think. And whether or not they are estranged from them, the families are hardly likely to be completely surprised by the homosexual transsexuals' decisions. More often, parents and siblings will react with "What took you so long?" As for work, homosexual transsexuals are less likely to hold conventional jobs, and those that do would have been recognized as being quite feminine and undoubtedly gay long before their transition.

In contrast, many autogynephilic transsexuals have both families and employers who will be shocked and disturbed by their decision. Although the autogynephile's wife often knows about his cross-dressing, she has typically discounted the possibility that this would lead to her husband's becoming a woman, often due to his assurance. (Early in their marriage, he probably doubted that this would occur, too.) Their children typically have no clue. Because the autogynephile is not usually outwardly feminine and has conducted his cross-dressing secretly, his coworkers and boss have probably never suspected anything either. For these men, there is no avoiding a crisis, one that usually causes profound alterations in their lives.

Some autogynephilic transsexuals would like to diminish the trauma of transition by easing into it. It is not uncommon for them to request gender clinics to allow them to gradually feminize their bodies, becoming increasingly androgynous, and change their female identities only after most people start treating them as women. Some actually attempt this—Petersen says that one such sign is an otherwise unremarkably masculine man who begins wearing clear nail polish. The Clarke Institute does not count such gender "blurring" toward the two years of real-life experience. The concern is that it avoids precisely the kind of information that transsexuals need—what it is like to live as a woman. Furthermore, Petersen thinks that transsexuals who try to adopt an ambiguous outward gender role might create more of a sense of discomfort or confusion among others than the actual transition would. Instead, she recommends that before transitioning, the transsexual should explicitly notify those who need to know what is going on. At work, this should be the boss first. "Bosses don't like hearing about this secondhand," says Petersen. Increasingly, employers are behaving sympathetically toward their transsexual employees. The most difficult situations are those in which the transsexual's pre-transition job required a great deal of interaction with the public (sales, for example). In this case, the employer might reasonably be concerned that the transsexual's continued employment in that position will cost the company business.

Marriages usually end. Individual wives' reactions vary from sympathetic and understanding to angry and hateful, but even in the best cases, women dislike the embarrassing notoriety and the loss of their husbands. (After all, they are not lesbians.) As in all divorces, the degree of animosity between parents is a major factor in how children come to view the noncustodial parent. Petersen thinks that it is important for the transsexual parent (ideally, but not necessarily in alliance with the other parent) to explain to the children before any transition, to emphasize that it was nothing they caused, and that the transsexual parent will continue to be a parent. This is an emotional issue for Petersen, because of her own experience. A postoperative transsexual, she has not seen her children since she transitioned socially into the female role in 1991. When Maxine Petersen was a man, on the day he planned to begin a slow process of talking to his children, gradually explaining the transition to them and getting them used to the idea, his then-wife talked to them first. Although he spent several hours with them afterwards, they were sufficiently traumatized that there was no hope of reaching them. Afterwards, the children told him they did not want to see him anymore. Petersen has called the children regularly and remembered birthday and Christmas presents, which she leaves at the front door. She also writes them occasional long letters. "I tell them that I love them and that the change has been only on the outside. On the inside, I am the same person who raised them, read them bedtime stories," she says, tearfully. I believe her.

★★★★★★★★★

Different nations range widely in the compassion and assistance given transsexuals. In the Netherlands the government pays for sex reassignment, even, in some cases, for adolescents. In Canada, the government used to pay, provided the applicant was treated through the Clarke Institute, but in 1998 the government ceased public funding. In England, transsexuals cannot currently legally change their sex, though they can get their medical expenses paid by national insurance. In Japan, sex reassignment surgery was not permitted until recently, when the first case (a female changing to male) was sanctioned. Islamic countries are especially intolerant. In Malaysia, for example, 45 contestants of a drag queen show were recently arrested for female impersonation; needless to say, sex reassignment surgery is not subsidized there. In the United States, of course, transsexuals can both obtain surgery and change their legal sex. However, private insurance almost never pays for the surgery, or for anything else involved in sex reassignment. Private insurance companies are motivated to keep costs to a minimum, and there are too few transsexuals to comprise a constituency to be reckoned with.

Almost certainly, refusal to cover sex reassignment surgery is also motivated by moral ambivalence. My undergraduate students at Northwestern are surely more liberal than average (at least until they get their first jobs or advanced degrees and begin to protect their assets), but even most of them balk at the idea that the surgery should be subsidized. They are especially hesitant to support surgery for nonhomosexual transsexuals, once they learn about autogynephilia. The idea of men sexually obsessed with having vaginas is incomprehensible to them, and like most Americans, they are too puritanical to give sexual concerns much priority in the public trough. But even when I invoke the standard transsexual narrative—"Imagine that you have felt your entire life that you had the body of the wrong sex"— they balk. When I press them, they say something like the following: "But they don't have the wrong body. They are mentally ill."

Paul McHugh, chairman of the Department of Psychiatry at Johns Hopkins University, used a more sophisticated version of that argument to close Hopkins's renowned gender identity clinic. McHugh objected that clinicians naively accepted transsexual patients' histories of having been quite feminine, when there was ample evidence in many cases that the histories were false (for example, a married man who presents as conventionally masculine). This objection is often correct, though it has no obvious relevance to the advisability of sex reassignment. Furthermore, and more importantly, McHugh argued that it is simply wrong for physicians to "mutilate" perfectly good organs because the transsexual patient's troubled mind wants this: "[The focus on surgery] has distracted effort from genuine investigations attempting to find out just what has gone wrong for these people—what has, by their testimony, given them years of torment and psychological distress and prompted them to accept these grim and disfiguring surgical procedures."

McHugh's concerns are worth taking seriously. Consider the case of the man erotically obsessed with having his leg amputated. Would it be advisable or even ethical to remove the leg? And McHugh is correct that interest in sex reassignment medicine has far exceeded interest in changing the minds of transsexual people so that they do not want to change their sex. Transsexualism is, after all, a condition of the mind and brain.

One problem with McHugh's analysis is that we simply have no idea how to make gender dysphoria go away. I suspect that both autogynephilic and homosexual gender dysphoria result from early and irreversible developmental processes in the brain. If so, learning more about the origins of transsexualism will not get us much closer to curing it. Given our present state of knowledge, saying that we should focus on removing transsexuals' desire to change sex is equivalent to saying that it is better that they should suffer permanently from gender dysphoria than that they obtain sex reassignment surgery.

Surely the most relevant data are transsexuals' own feelings before and after transitioning. Are they glad they did it? By now, hundreds of transsexuals have been followed after changing sex, and the results are clear. Successful outcomes are much more common than unsuccessful outcomes. In the typical study perhaps 80 percent of male-to-female outcomes are judged successful, about 10 percent unsuccessful, and about 10 percent uncertain. (The results of genetic females who become men are even more successful.) "Success" has been defined differently by different investigators, and has included such things as absence of regrets, and success in work and sexual relationships. No one claimed that transsexuals were without problems, only that they seemed to have adjusted well. Furthermore, the few studies that had adequate control groups found that as transition progressed through hormones and then surgery, patients' well-being also increased and surpassed that of those waiting sex reassignment.

Those patients who did have regrets tended to have had poor surgical outcomes, work-related problems (for instance, dismissal because of transitioning), or poor functioning to begin with. There was also some indication in a couple of studies that autogynephiles were more likely to have regrets. In a late-1980s study by Blanchard, about a third of a small sample of nonhomosexual transsexuals had some regrets. Both Blanchard and Petersen believe that the regrets rate among autogynephiles would be much lower now because of greater tolerance among employers and a more cautious approach to recommending patients for surgery.

As vaginoplasty has become more and more sophisticated, transsexual patients undoubtedly have been increasingly satisfied. Juanita was initially unhappy with the look of her neo-vagina after phase one surgery, but after the second phase, she was delighted. She had been insecure that her partners would detect that her vagina was not real before; now she doesn't worry. In the past, neo-vaginas tended to be shallow. In one study from the 1980s, they averaged about three inches, and the researchers speculated that transsexuals who had sex with men might use sexual positions that minimize depth of penetration. With the recent trend of using skin grafts to lengthen the vagina, this is becoming less of a concern, if it ever was one. None of the transsexuals I met complained about vaginal depth. Some of them—particularly the homosexual transsexuals—had been concerned before sex reassignment surgery that they would lose erotic sensation and become anorgasmic. And this does occur, though most transsexuals retain the capacity for orgasm. In fact, several transsexuals claimed to me that for the first time in their lives, they were experiencing multiple orgasms. Blanchard is skeptical about such accounts, because he suspects they are trying to convince themselves or others that they are like genetic women. However, they have convinced me that there is something really going on. Do they have multiple orgasms just like some genetic women do? I am not sure, because as a man I cannot understand the phenomenon; nor has it been well understood scientifically.

Although there have been a number of follow-up studies of transsexuals, these studies have been quite limited in terms of their outcome measures, among other things. Because of this, perhaps the best indication that sex reassignment is usually successful is that transsexuals continue to seek it. They do not seek it blindly. On the contrary, as I have mentioned, transsexuals are highly motivated and educated consumers of sex reassignment. Many know the scientific literature, and all of them closely question other transsexuals at more advanced stages of transition. If, from their perspective, sex reassignment were a bad idea—if transitioning routinely led to unhappiness—they would not go through with it.

Of course, there are other perspectives that deserve to be weighed. Most obviously, the wives and children of autogynephilic transsexuals might well be less happy after their husbands and fathers change their sex. I think their suffering is understandable and unfortunate. (I am less sympathetic toward disapproving families of origin of homosexual transsexuals, who do not depend on them.) However, I do not think that this real suffering should be used to discourage transsexuals from sex reassignment. Most are plenty conscious of the suffering they might cause their families and proceed, if they do, with regret about this. And in a society in which nearly half of marriages end in divorce, often caused primarily by boredom, it is difficult to understand why autogynephilia is not sufficient reason to end a marriage.

****

Do transsexuals find partners? Certainly, homosexual transsexuals find sex partners after their surgery, but do they get *steady* partners? Do they get married? I have already mentioned my impression that homosexual transsexuals are not very successful at finding desirable men willing to commit to them. In part, this reflects the difficulty that men have with the notion of coupling with women who used to be men (no matter how attractive such women may be), as well as the difficulty most transsexuals have keeping their secret. But it also reflects the choices that homosexual transsexuals are prone to make. My impression is that they would rather have a relatively uncommitted relationship with a very attractive man than a committed relationship with a less desirable partner. Although the homosexual transsexuals I have met are all searching for "Mr. Right," perhaps in vain, their sex lives have all clearly improved after surgery. They can hide their past identities for a while, at least, and so no longer have to worry about how to respond to attractive men who hit on them in bars.

When I began writing this book, I had never known a homosexual transsexual who married. However, in 1999 Juanita invited me to her wedding. Her engagement story was quite romantic, in an odd, transsexual kind of way. She met her fiancé on the Internet a couple of years after her vaginoplasty. When they began dating, she didn't tell him her secret. They were on vacation in Mexico, and he proposed. She began her answer with, "There's something I've been meaning to tell you." After she confessed, he was stunned, but he told her he wanted to marry her anyway.

The wedding was small, touching, and hilarious. Juanita's family—mother, father, three brothers, and three sisters—all attended, and of course, they knew that Juanita used to be Hector. As did the four transsexuals, including Cher, whom Juanita invited. However, neither the groom's parents nor his son from his first marriage had any idea. Juanita was radiant, but when I spoke privately with her, she revealed that she was having second thoughts about becoming stepmother to a teenage boy and living in the suburbs. But she reminded herself what a great catch her husband was.

Just a year later Cher told me that Juanita and her husband were separated. Apparently, Juanita's doubts had only grown, and she missed the excitement of living in the city, and of dating new partners. She had also begun to work again as an escort—she had done this before meeting her husband. Juanita had achieved the dream that nearly all the homosexual transsexuals I've met have told me they want, and she let it go.

Nearly all the homosexual transsexuals I know work as escorts after they have their surgery. I used to think that somehow, they had no other choice because conventionally happy lives were beyond their grasp. I have come to believe that these transsexuals are less constrained by their secret pasts than by their own desires. And these desires, including the desire for sex with different attractive men, do not make conventional married life easier.

Autogynephilic transsexuals tend to lead very different sex lives than homosexual transsexuals, both before and after surgery. Autogynephiles are more likely to seek one single partner. A few remain with their wives, though much more often, wives divorce them. A significant number of autogynephiles find lesbian partners. It is not uncommon for autogynephilic transsexuals to pair up with each other. My impression is that a substantial proportion of autogynephilic transsexuals do not get partners (even casual sex partners) after their surgery. However, this doesn't mean that in these cases sex reassignment surgery has failed. Autogynephilic transsexuals do not primarily seek sex reassignment in order to attract partners.

****

Cher has been having a rough time lately. She has fallen out with Amy, a homosexual transsexual who used to be her closest friend. Cher thinks that once Amy got her surgery, she no longer needed her, and she feels used. When she goes out with Juanita, who has become her best friend, men are constantly approaching Juanita (who is 15 years younger and very sexy), but they approach Cher cautiously, if at all. Cher also admits that she is strongly attracted to both Amy and Juanita (and I wonder if she has fallen in love with them). Of course, they have no romantic or sexual interest in her, or for anyone who is not a man, and so her lust is unrequited. Cher sounds depressed sometimes and worries that she will never find anyone. She is also broke, and is being sued by her relatives for her father's inheritance.

I ask her if she ever regrets becoming a woman, and she does not hesitate. "No, that is one thing I know was right," she says. "I do not regret that, and I am not ashamed of anything."

Despite her troubles, she continues to visit her circle of (primarily transsexual) friends, helping them plan their transition, listening to their boyfriend problems, and urging them away from those areas of transsexual life of which she disapproves—prostitution, for instance. She is a good friend to them, although her advice is not always appreciated or heeded.

I think about what an unusual life she has led, and what an unusual person she is. How difficult it must have been for her to figure out her sexuality and what she wanted to do with it. I think about all the barriers she broke, and all the meanness that she must still contend with. Despite this, she is still out there giving her friends advice and comfort, and trying to find love. And I think that in her own way, Cher is a star.

Epilogue

I recently met Danny Ryan for the first time. Although I had discussed him several times with his mother, Danny had steadfastly refused to meet me. I believe that he associated me with concern about his femininity, and he didn't like to be reminded of this.

I met him by accident. I attended my university's graduation ceremony, and Danny's family did as well. His parents—Leslie and Patrick Ryan—Danny, and his sister were there in support of a family friend's daughter. I was there because it was my turn to represent our faculty at convocation. After the ceremony, I passed them in a long hallway, and Leslie and I simultaneously noticed each other.

As Leslie Ryan introduced me to her family, I could not help focusing my attention on Danny. A slender boy with medium-length (for a boy) light blonde hair, blue eyes, and fine features, he was impeccably dressed in a navy suit, red tie, and black shoes. As Leslie introduced him to me, I tried to notice his response, to see whether he recognized my name. I couldn't tell.

Looking at Danny, it was difficult to imagine him wearing high heels and a dress. He looked good as a boy—if an unusually formally dressed one. When the family friend's daughter showed up, she told him how handsome he looked, and he beamed. This was not a girl in boy's clothing.

As we congregated in the hallway, I watched Danny interact. Shy at first, he whispered quietly to his sister. Then someone asked him about Convocation. He cocked his head back dramatically, threw his forearm across his eyes and said, "I thought it was entirely too long. Must they read every single name?" His word choice was obviously unusual, for an eight-year-old boy, and his speech style was precise and somewhat prissy. This was not a typical boy, either.

A few moments later, Danny said: "Mummy, I need to go to the men's room." I am certain that as he said that, he emphasized "men's" and looked my way. And off he went, by himself. At that moment, I became as certain as I can be of Danny's future.

Further Reading

CHAPTER 1

The place to start is not a book but a film. The Belgian film *Ma Vie en Rose* (*My Life in Pink*, Alain Berliner, Sony Pictures Classics, 1997) is a startlingly effective portrayal of a very feminine boy. The story is poignant and entertaining. See it.

CHAPTER 2

Richard Green's classic book, *The 'Sissy Boy Syndrome' and the Development of Homosexuality* (New Haven: Yale University Press, 1987) is that rare combination of an important scholarly contribution and a good read. For a more exhaustive (and academic) treatment, see *Gender Identity Disorder and Psychosexual Problems in Children and Adolescents* (New York: Guilford Press, 1995) by Kenneth J. Zucker and Susan J.

Bradley. For the conservative, anti-gay approach to atypical gender identity, see George Rekers' writings on the National Association for Research and Therapy of Homosexuality website (www.narth.com). For the far-left approach, which criticizes even moderates like Zucker, read *Gender Shock* (New York: Anchor, 1997) by Phyllis Burke.

CHAPTER 3

As Nature Made Him (New York: Harper Perennial, 2001), by John Colapinto, focuses on the famous case of David Reimer, who lost his penis in infancy and was raised unsuccessfully as a girl. *Man and Woman, Boy and Girl* (Baltimore: Johns Hopkins University Press, 1973) is the classic book by John William Money and Anke A. Ehrhardt that summarizes the research from the 1950s and 1960s that was used to justify the decision to reassign Reimer as a girl.

CHAPTER 4

The Naked Civil Servant (New York: Penguin, 1997), by Quentin Crisp, is a wonderful book about gay femininity. It is the autobiography of Crisp, who is the prototype of the queenie gay man. (In typical British fashion, Crisp was not as loud and dramatic as the stereotypic American queen, but he otherwise fit the bill.) It is both hilarious and insightful.

CHAPTER 5

Several books illuminate the world of gay casual sex. *Tearoom Trade* (New York: Aldine de Gruyter, 1975), by Laud Humphreys, is a study of men who engage in homosexual acts with strangers in public restrooms. Larry Kramer's controversial book, *Faggots* (New York: Grove Press, 2000), gives a harsh portrait of sexual excess among gay men during the 1970s. *And the Band Played On* (New York: St. Martin's

Press, 2000), by Randy Shilts, is about the AIDS epidemic, a tragic consequence of gay promiscuity.

CHAPTER 6

Simon LeVay's *Queer Science* (Cambridge, Massachussetts: MIT Press, 1997) is an excellent introduction to the biology of sexual orientation. For genetics, Dean Hamer and Peter Copeland's *The Science of Desire* (New York: Touchstone Books, 1996) is a fine introduction. Whether or not Hamer's Xq28 finding turns out to be correct, his general knowledge about sexual orientation is excellent, and he is witty and refreshingly outspoken.

CHAPTER 7

The clearest exposition of the social constructionist case is Edward Stein's *The Mismeasure of Desire* (New York: Oxford University Press, 2001). The most influential books have included *One Hundred Years of Homosexuality* by David M. Halperin (New York: Routledge, 1989) and Michel Foucault's opus *The History of Sexuality* (New York: Vintage, 1990). But be warned that neither of these is easy (or in my opinion, clear) to read. The historian Rictor Norton has written the most thorough critique of social constructionist accounts of homosexuality: *The Myth of the Modern Homosexual* (London: Cassell Academic, 1998).

CHAPTERS 8 AND 9

The current popular literature on transsexualism is noteworthy in its ignorance of the distinction between autogynephilic and homosexual transsexuals. The biographies and autobiographies of famous transsexuals, including those of Christine Jorgensen (*The Christine Jorgensen Story*, Susan Stryker and Christine Jorgensen, Chicago: Cleis

Press, 2000) and Deirdre N. McCloskey (*Crossing: A Memoir,* Chicago: University of Chicago Press, 1999), focus on the standard transsexual story ("I always felt like a female"), even though I would guess that both Jorgenson and McCloskey were autogynephilic. This is also true of books about transsexualism and sex reassignment, such as *True Selves* (Mildred L. Brown and Chloe Ann Rounsley, San Francisco: Jossey-Bass, 1996). Some exceptions include Richard Ekins' *Male Femaling* (New York: Routledge, 1997), which provides an excellent sociological overview.

The book *Transvestites* (Buffalo, NY: Prometheus Press, 1991) shows why Magnus Hirschfeld is a giant of twentieth-century sexology. Look at case #12, especially, to see that Hirschfeld understands something about autogynephilia, long before Blanchard nailed the concept down.

Ray Blanchard has, unfortunately, written only for academic audiences. His most accessible treatment of autogynephilia might be "Clinical Observations and Systematic Studies of Autogynephilia," published in 1991 in the *Journal of Sex and Marital Therapy* (*17(4)*: 235-251).

Anne Lawrence maintains an awesome website for transsexuals, Transsexual Women's Resources, (www.annelawrence.com/twr), and one section of her site is devoted to autogynephilia. (See the "Sexuality" link.) Not only does she have clear explanations of autogynephilia, but she also includes testimonials of transsexuals who have visited her site and read about the concept. Most of them are thankful that someone is finally talking about the sexual side of transsexualism. Some say they finally understand themselves. A few are angry with Anne for embracing Blanchard's "wrongheaded" ideas. Anne's essays on autogynephilia have been translated into French, Italian, and Chinese by grateful readers.

For an article that angered many autogynephiles—but which provides a sympathetic portrayal of both cross-dressers and their wives— see Amy Bloom's "Conservative Men, Conservative Dresses," originally published in the *Atlantic Monthly*, April 2002, and included in her collection *Normal* (New York: Random House, 2002).

CHAPTER 10

The documentary film, *Paris Is Burning* (Jennie Livingston, Miramax/NEA, 1990), provides a sympathetic and accurate picture of mostly African-American transsexuals and drag queens in New York City. *Honey, Honey, Miss Thang* (Philadelphia: Temple University Press, 1996), by Leon E. Pettiway, is a book treatment of a similar culture. The film *Wigstock* (Barry Shils, 1995) focuses more on drag performance than on life between performances, but it, too, is worth watching. *The Queen* (Frank Simon, 1968) is an earlier documentary of a drag contest that has several very funny moments along with some poignant ones.

CHAPTER 11

If you are considering a sex change from man to woman, the best place to start is Anne Lawrence's website (www.annelawrence.com/twr), where you will get up-to-date information about procedures used by different surgeons, as well as photographs of their work. She also has sections on other relevant issues, such as voice feminization and outcome research. As far as books go, *Miss Vera's Finishing School for Boys Who Want to Be Girls* (New York: Doubleday, 1997), by Veronica Vera, is instructive.

Index

𝒜

Abbé de Choisy, 171
Abortion controversy, 114
Adoptive brothers, 108
AIDS/HIV infection, 84, 86, 108,
110, 119, 120, 121, 162,
183, 185, 217
Allen, Laura, 119, 120
Ambady, Nalini, 73–74
American Psychiatric Association,
81
Amputee fetish, 201, 206
Anal sex, 57, 83–84, 128, 131, 148,
156, 188
Anderson, Clinton, 26
Anxiety, 81–82
Apostle Paul, 128
Archives of General Psychiatry, 110
Aristophanes, 128
Artistic brother hypothesis, 117–118
Autoerotic asphyxia, 172

ℬ

Bahuchara Mata, 134
Bawer, Bruce, 100
Behavior modification therapy, 24–
26, 28
Bend Over Boyfriend (video), 83
Benjamin, Harry, 174, 201
Bergling, Tim, 59
Bestiality, 171
Birdcage, The (movie), 76, 77
Birth order hypothesis, 111, 158
Bisexuals and bisexuality, 95, 107,
108, 132, 133, 136, 159,
162
Blanchard, Ray, 157–159, 162,
165–167, 168, 169, 176,
186, 192, 201–202, 207–
208, 218
Bloom, Amy, 218–219
Boston University, 106
Boswell, John, 128

Brain development
    and gender identity
        development, 44, 207
    INAH-1 and INAH-3, 119–
        122
    sexual differentiation in, 72–
        73, 118–122, 169
    Sexually Dimorphic Nucleus
        (rats), 119–120
Breedlove, Marc, 123
British public schools, 125, 131,
    132
Brown, John Ronald, 200
Burke, Phyllis, 26, 27
Byne, Bill, 121–122

C

Canada, 205
Casual sex, 86, 87–90, 101, 156,
    185, 211, 216–217
Catholic priests, 66, 69
Centers for Disease Control and
    Prevention, 86
Chase, Cheryl, 174
Chicago
    Baton, 186–187
    gay district, 59–60, 63–64, 96,
        128, 141, 142
Childhood behavior
    autogynephilic cross dressing,
        170
    awareness of unacceptability
        of, 7–10
    with cloacal exstropy, 41–42
    feminine boys, x, 3–15, 18, 68,
        147

male homosexuals, x–xi, 12,
    13, 17, 19–20, 38, 58–59,
    61–62, 68, 80–81
    sexual knowledge, 34–35
Childrearing practices. *See*
    Parenting behavior
Clark Institute of Psychiatry, 157,
    162, 163, 170, 172–173,
    201, 204, 205
Cloacal exstrophy
    assessment of outcomes, 49–50
    childhood behavior, 41–42
    female assignment of males,
        40–44, 47
    and gender identity
        development, 42–45, 47,
        48–50, 51–52
    male assignment of males, 51
    and persecution and
        stigmatization, 46
    prevalence, 41
    refusal of sex reassignment, 49
    sex reassignment surgery, 40–
        44, 47, 49, 51
    and sexual orientation, 45, 46,
        47, 48, 111
    telling children about, 42–43,
        51
Cohen-Kettenis, Peggy, 32
Corbett, Ken, 26
Creation myth, 128
Crisp, Quentin, 79, 84, 216
Cross–dressing/cross–dressers
    autoerotic asphyxia and, 172
    in childhood, 3, 4, 6, 18, 147,
        170

conditioning explanations, 169
counseling for, 164
by drag queens, 135, 159–160, 179–180
erotic component to, 152–153, 157, 160–161, 163–164, 170, 173, 174–175, 176
genetic link, 170, 172
guilt and purging, 153, 161, 171
heterosexual transvestites, 155, 160–161, 165
occupational, 135
sexual sadism and, 172
she-male prostitute customers, 187–188
stigmatization, 161
by transsexuals, 147, 148, 152–153, 155, 159, 163–165, 168, 169–170, 173, 203, 218–219
Crying Game, The (movie), 79, 142, 198–199
Cultural influences
feminine boys, 73
sexual behavior of male homosexuals, 133–138
and transsexualism, 171, 183–184

D

Dancers, 67–69, 76
Davidson, Jaye, 79
Denny, Dallas, 200–201

Depression, 81–82, 83, 155, 163
Die Transvestiten (Hirschfeld), 171
Dörner, Gunter, 104–105
Dominatrix, 188
Drag queens, 72, 76, 135, 136, 155, 159–160, 179–180, 183, 205, 219

E

Eating disorders, 82, 93
Ed Wood (movie), 142
Egalitarian relationships, 134, 137–138
Estrogen therapy, 46, 48, 149, 196–197
Evolution
and age preferences in mates, 96–97
homosexuality as a paradox, 88, 115–118
and interest in casual sex, 87–88
and visual sexual stimuli, 94
Exhibitionism, 171

F

Family relationships
feminine boys, 18–19, 26, 30, 37–38, 103
gay sons, 37, 98–99, 114–115
of transsexuals, 15, 20, 32, 149–150, 155–156, 183, 203, 204–205, 206, 209, 211
Fathers, and feminine sons, x, 4, 6, 17, 19–20, 26, 37, 103

Female impersonators, 185, 186–187, 205

Feminine boys
    behavior modification therapy, 24–26, 28
    causes attributed to, x, 10–12, 22–34, 39, 52–54, 111, 169; *see also* Gender identity disorder
    childhood behavior and experiences, x, 3–15, 18, 68, 147
    childrearing practices and, 10–12, 16, 18–19, 20–22, 30, 31, 39, 52, 103
    cross dressing, 3, 4, 6, 18, 147
    cultural influences, 73
    and family dynamics, 18–19, 26, 30, 37–38, 103
    friendships, 8–9
    genetic link, 11–13, 22, 82–83, 103–104, 108, 118
    Green's study, 17–19, 25–26
    normalization of gender role behavior, 184, 213–214
    occupational and recreational interests, 65, 68–69
    prenatal effects of hormones, 53–54
    psychological treatment, 15, 20, 22, 23–28, 29, 30–31, 38
    sexual orientation, x–xi, 12–13, 17, 19–20, 26, 36, 38, 58–59, 61–84, 91, 93, 108, 111, 134, 135–136, 138
    shame and defensiveness, 15, 37, 58–59, 103
    speech development, 73, 214
    stereotypes, 17
    stigmatization and persecution, xi, 7–10, 14–15, 17, 30, 31, 32–33, 35–36, 38, 80, 183
    suicide, 37
    and transsexualism, xi, 15, 20, 28, 30, 31–32, 33, 135–136, 138, 146, 148, 163, 168, 178, 183, 184

"Femmes," xi, 36, 78, 79, 80, 84, 93, 109

"Femiphobia/sissyphobia," 59, 77–78, 80

Filipino *bayot*, 136

Finger length ratio, 122, 123

Florence, Italy, in fifteenth-century, 125, 129

Frotteurism, 171

Freund, Kurt, 95, 157, 158

G

Gay men. *See also* Male homosexuality
    accent/speech, xi, 69–73, 75–76, 79, 79–80
    age preferences in mates, 96–98, 131, 132
    AIDS/HIV infection, 84, 162, 183
    childhood behavior, x–xi, 12, 13, 17, 19–20, 38, 58–59, 61–62, 68, 80–81

coming out, 19, 36, 37, 57

drag queens, 72, 76, 135, 136, 155, 159-160, 179-180, 183, 205

earliest sexual experiences, 36, 112, 148

"femmes," xi, 36, 78, 79, 80, 84, 93, 109

"femiphobia/sissyphobia," 59, 77-78, 80

flaming, 79

knowledge of sexuality, 35-36

mating psychology and relationships, 76-81, 87, 90-91, 98, 100-101

monogamy, 85-86, 90, 100-101

movement patterns and motor behavior, xi, 73-76, 79-80

number of sex partners, 86, 100

occupational and recreational interests, 63-69, 79-80, 105-106

persecution by other gays, 35-36

personals ads, 77-79, 84, 86-87, 93, 98, 101

prostitutes/prostitution, 134, 135, 136

psychological problems, 81-83, 93

research as a threat to, 113-115, 158

sex with women, 57, 116

sexual practices and preferences, 57, 83-84, 86

stereotypes, 63-81

stigmatization/social intolerance, 37, 81, 100, 114, 115

twins, 106-110

Gay marriage, 100

Gay rights movement, 100

Gender dysphoria, 23, 27

Gender identity, xi, 22, 47
  defined, 50
  gradations in, 50
  "look at 'em funny" theories of, 52-53

Gender Identity and Psychosexual Disorders Clinic, 48

Gender identity development
  brain differences, 44
  cloacal exstrophy and, 42-45, 47, 48-50, 51-52
  hormones and, 42, 44, 53-54
  nature-nurture debate, 44-47, 52-53
  psychosexual neutrality at birth, theory of, 47-48
  socialization and, 39-40, 41-44, 52, 65

Gender identity disorder
  behavioral criteria, 22-23
  conservatives' theories and therapies, 24-26, 31
  controversies, 23-24
  defined, 22
  liberals' theories and therapies, 26-28, 29-30, 31
  moderates' theories and therapies, 28-34

research design, 34
transsexual outcomes, 28, 31–
   32, 33–34, 167, 178–179,
   180
*Gender Identity Disorder and
   Psychosocial Problems in
   Children and Adolescents*
   (Zucker), 29, 215
Gender role behavior, xi, 45, 184
Gender Shock (Burke), 26, 216
Genetic testing for homosexuality,
   113–114
Genetics
   and cross dressing, 170, 172
   and femininity in males, 11–
      13, 22, 82–83, 103–104,
      108, 118
   multiple gene hypothesis, 118
   of sexual orientation, 58, 106–
      110, 112–113, 118
   twin studies of gays, 106–110
   X-linked trait, 112–113
Genital arousal, measurement of,
   94
Greek pederasty, 124, 125, 126,
   127, 128, 131
Green, Richard, 17–19, 25–26, 32,
   62, 178, 215
Gynandromorphophiles, 186

H
Hamer, Dean, 112, 120, 121
Harry Benjamin International
   Gender Dysphoria
   Association, 201
Harvard Sentences, 70

Harvard University, 24
Heimel, Cynthia, 93
Heterosexual outlets, and male
   homosexuality, 131–132
Hirschfeld, Magnus, 171
Homosexuality. *See also* Gay men;
   Male homosexuality
   abortion controversy, 114
   artistic brother hypothesis,
      117–118
   children's knowledge of, 34–
      35, 36
   creation myth, 128
   genetic testing for, 113–114
   inclusive fitness hypothesis,
      117
   parents' reaction to, 98–99,
      114–115
   as perversion or inferior
      sexuality, 24, 101
   prevalence of, 111–112, 126–
      127
   psychological therapy, 29
   removal from *DSM*, 81
   as social construction, 124–133
   as solution to overpopulation,
      117
Hooker, Evelyn, 81

I
Immune response hypothesis, 111
Inclusive fitness hypothesis, 117
Indian *hijras*, 134, 138
Intercrural intercourse, 127
International Academy of Sex
   Research, 158

Intersexuality, 175
Isay, Richard, 26, 27

J

Japan, 205
Johns Hopkins University, 47, 48, 206-207
Jorgensen, Christine, 20, 217

K

Kenrick, Doug, 98
Kleinfelter's syndrome, 22

L

Lawrence, Anne, 168, 174-175, 218, 219
Leadership U, 24
Leno, Jay, 123Lesbians, 47, 67-68, 99, 100, 122-123, 185, 204, 211
LeVay, Simon, 26, 27, 28, 32, 119-121, 217
Lippa, Richard, 64-65, 66, 123

M

Ma Vie en Rose (film), 215
Machiavelli, 129
Malaysia, 205
Male homosexuality. *See also* Gay men
  in adoptive brothers, 108
  brain development and, 72-73, 118-122, 169
  birth order and, 111, 158
  in British public schools, 125, 131, 132
  cultural differences in behavior, 133-138
  egalitarian relationships, 134, 137-138
  evolutionary paradox, 88, 115-118
  family acceptance of, 37
  and femininity, x-xi, 12-13, 17, 19-20, 26, 36, 38, 58-59, 61-84, 91, 93, 108, 111, 134, 135-136, 138
  in fifteenth-century Florence, 125, 129
  genetic link, 106-110, 112-113
  Greek pederasty, 124, 125, 126, 127, 128, 131
  heterosexual outlets and, 131-132
  immune response hypothesis, 111
  Indian *hijras*, 134, 138
  and masculinity, 58, 60, 84, 85-102, 108, 135-136
  Native American *berdache* tradition, 134, 137, 138
  Oman *xanith*, 135
  in prison, 131
  Roman cinaedi, 125, 128
  in Sambia tribe of New Guinea, 125, 130-131, 132
  seduction-and-recruitment hypothesis, 112
  self-awareness of, 36-37
  shame and conflict about, 26, 27, 31, 33, 36, 59, 108

Tahitian *mahu*, 134
transsexual, 134, 135, 136,
    137, 138, 146, 147–151,
    159, 162, 163, 167, 168–
    169, 171, 172, 176, 177–
    191
    in twins, 106–111
    as X-linked trait, 112–113
Marks, Mimi, 186
Masculine females, xii
Masculinity, and male
    homosexuality, 58, 60, 84,
    85–102, 108, 135–136
Masochism, 171, 172
Mating psychology and
    relationships of gay men,
    76–81, 87, 90–91, 98,
    100–101
Mattison, Andrew, 90
McCloskey, Deirdre N., 218
McHugh, Paul, 206–207
McWhirter, David, 90
*Midnight in the Garden of Good and
    Evil* (movie), 143
Money, John, 47–48, 50, 155, 216
Monogamy, 85–86, 90, 100–101
Mothers, and feminine sons, 3–5,
    8–15, 18–19, 20–21, 26,
    37, 39, 52
Movement patterns and motor
    behavior, xi, 73–76, 79–80
Mutual masterbation, 131

$\mathcal{N}$

Naked Civil Servant, The (Crisp),
    84, 216

National Association for Research
    and Therapy of
    Homosexuality
    (NARTH), 24, 216
National Institutes of Health, 24
Native American *berdache* tradition,
    134, 137, 138
Nature (journal), 122–123
Necrophilia, 171
Netherlands, 32, 82, 197, 205
Neuroticism, 82–83
Northwestern University, 16, 64,
    86, 107, 206
Norton, Rictor, 217

O

Oberschneider, Michael, 67
Occupational and recreational
    interests
    of feminine males, 65, 68–69
    of gay men, 63–69, 79–80,
        105–106
    of transsexuals, 136, 142, 149,
        152–153, 168, 184–185,
        210
Oman *xanith*, 135
Ontario Correctional Institute, 157
Oral sex, 57, 86, 130, 131, 134,
    148, 149, 188

$\mathcal{P}$

Paraphilia, 171–172
Parenting behavior
    and femininity in boys, 10–12,
        16, 18–19, 20–22, 30, 31,
        39, 52, 103

study design, 21-22
Pederastic relationships, 128
Pedophilia, 171
Penile ablation, and gender
    development, 45-47
Penile plethysmograph, 94, 95
Persecution and stigmatization
    cloacal exstrophy and, 46
    cross dressing and, 161
    of feminine boys, xi, 7-10, 14-
        15, 17, 30, 31, 32-33, 35-
        36, 38, 80, 183
    of gays by other gays, 35-36
    of homosexuals by
        heterosexuals, 37, 81, 100,
        114, 115
    of transsexuals, 183
Personals ads, 77-79, 84, 86-87, 93,
    98, 101
Peterson, Maxine, 172-173, 202,
    203, 204-205, 208
Pillard, Richard, 106, 110, 121
Playgirl magazine, 94
Pornography, 93-96
Prejudice against "femmes," xi
Prenatal effects of hormones, 53-
    54, 104-105, 119-120,
    121, 122
Pressland, Natasha, 135
Priscilla: Queen of the Desert
    (movie), 142
Prisons, homosexuality in, 131
Prostitutes/prostitution, 134, 135,
    136, 151, 183-185, 187-
    188, 211
Pseudo-hermaphrodites, 47

Psychological problems, of gay
    men, 81-83, 93
Psychosexual neutrality at birth,
    theory of, 47-48
Psychotherapy
    for cross dressing, 164
    for feminine boys, 15, 20, 22,
        23-28, 29, 30-31, 38
    for homosexuals, 29

R

Reimer, David (John/Joan case),
    45-46, 48, 51, 216
Reiner, William, 48, 51-52
Rekers, George, 24-25, 27, 28, 31,
    216
Richardson, Justin, 26
Rocke, Michael, 129
Rodman, Dennis, 142
Roman cinaedi, 125, 128

S

Sambia tribe of New Guinea, 125,
    130-131, 132
Science magazine, 120
Sedaris, David, 35, 73
Seduction-and-recruitment
    hypothesis, 112
Sex differences
    in age preferences in mates,
        96-98
    in brain development, 72-73,
        118-122
    in casual sex interests, 86, 88-
        90, 101
    finger length ratio, 122, 123

in height, 65
in interest in children, 99–100
in mating psychology, 76–77,
    87–88
in movement and motor
    behavior, xi, 73–76
in sexual arousal patterns, 94–
    96
in socialization, 65–66
in speech, 72–73
in visual sexual stimuli, 93–96
Sex reassignment surgery
age factors, 46–47
appropriateness for
    autogynephiles, 173, 174
breast implants, 198–199
castration, 196
cloacal exstrophy and, 40–44,
    47, 49, 51
costs, 149, 197, 198, 199
electrolysis, 196
employers' tolerance of, 203,
    208
facial plastic surgery, 198
in homosexual transsexuals,
    159, 182–183, 185, 188–
    191
hormone therapy, 46, 48, 149,
    196–197, 201, 202
insurance coverage, 205–206
international differences in
    compassion and
    assistance, 205–206
labiaplasty, 200
and orgasm, 208
penile ablation and, 45–47

penile construction, 45
postoperative regret, 173, 202,
    207–208
procedures, 195–201, 219
quack surgeons, 200–201
realism and sensitivity of neo-
    vagina, 195, 200, 208
and sexual relationships, 46,
    149, 185, 188–191, 209–
    211
side effects, 31, 197, 199
silicon injections, 199
social transition prior to, 201–
    205
successful outcomes, 207–209
vaginoplasty, 45–46, 189, 199–
    201
voice surgery or therapy, 197–
    198, 219
Sexual arousal patterns, 94–96
Sexual orientation, xi
of autogynephilic transsexuals,
    180, 183, 190–191
biological markers of, 122–
    123
cloacal exstrophy and, 45, 46,
    47, 48, 111
and dance, 67–69, 76
environmental factors, 110–
    112
essentialists' view, 126
of feminine boys, x–xi, 12, 13,
    17, 19–20, 38, 58–59, 61–
    63, 84
genetics of, 58, 106–110, 112–
    113, 118

of homosexual transsexuals,
136, 146, 151, 159, 182–
183
and interest in parenting, 99–
100
negative effects of research on
gays, 113–115
parental influence, 4
and physical attractiveness of
partners, 91–93
and physical violence, 91
prenatal effects of androgens,
53–54, 104–105, 119–120,
122
prenatal stress and, 53–54,
104–106
psychological hypothesis, 65–
66
and Rorschack test responses,
81
sexual arousal patterns and,
94–96
and sexual jealousy, 90–91
and sexual promiscuity, 86–87,
101
social constructionists' view,
124–133
sociological hypothesis, 66
of transsexuals, xi–xii, 136,
148, 152–154, 156, 178,
211
twin studies, 106–110
and visual sexual stimuli, 93–
96
Sexual partners, number of, 86,
100; *see also* Casual sex

Sexual sadism, 171, 172
Sexually Dimorphic Nucleus
(rats), 119–120
She-males, 186–188, 190
Siblings, and feminine brothers, 5
Silence of the Lambs (movie), 143
Social conservatives' theories and
therapies, 24–26, 31, 87,
89–90, 100, 111
Social constructionism controversy,
124–133, 217
Social liberals' theories and
therapies, 26–28, 29–30,
31
Social moderates' theories and
therapies, 28–34
Socialization, and gender
development, 39–40, 41–
44, 52, 65
Society for the Second Self (Tri-
Ess), 161, 168
Standards of Care for Gender
Identity Disorders, 201
Stereotypes, gay, 63–81
Suicide, 37, 83
Sullivan, Andrew, 100
Symons, Donald, 90
Symposium (Plato), 128

T

Tahitian *mahu*, 134
Testosterone, 42, 44, 53, 118–120,
122, 201
Thailand, 135
Tonga, 135
Transgendered, 174

Transexual Women's Resources, 218, 219

Transsexualism, female-to-male, 201, 205, 207

Transsexualism, male-to-female
  causes, 145, 166–167, 169–170, 171–172, 178, 206, 207
  defined, 144
  degrees of, 144
  and femininity, xi, 15, 20, 28, 30, 31–32, 33, 135–136, 138, 146, 148, 163, 167, 168, 178, 181, 184
  gender identity disorder and, 28, 31–32, 33–34, 167, 178–179, 180, 207
  genetic links, 183–184
  motivation for, xi–xii, 146, 164
  movies about, 142–143
  as paraphilia, 171–172
  prevalence, 20, 142
  surgery, *see* Sex reassignment surgery
  talk-show themes, 142–143
  therapy, 153, 162–163, 165–166
  types, 162–166

Transsexuals
  asexual, 159, 162, 163
  autogynephilic, 146, 147, 151–156, 163–176, 181, 190–191, 192–194, 202, 203–204, 206, 207–208, 209, 211, 218
  beauty pageants, 135

bisexual, 159, 162
  and casual sex, 156, 185, 211
  common lies and deceptiveness of, 172–176
  cross-cultural occurence, 171, 183–184
  cross dressing, 147, 148, 152–153, 155, 159, 163–165, 168, 169, 173, 174–175, 176, 203
  depression, 155, 163
  disclosure to partners, 190–191
  discrimination against, 183
  distinguishing between types, 192–194
  dreams for the future, 186, 210
  family relationships, 15, 20, 32, 149–150, 155–156, 183, 203, 204–205, 206, 209, 211
  female impersonators, 185, 186–187
  Filipino *bayot*, 136
  heterosexual, 159, 162
  HIV infection, 185
  homosexual, 134, 135, 136, 137, 138, 146, 147–151, 159, 162, 163, 167, 168–169, 171, 172, 176, 177–191, 207, 209–210
  marriage and children, 162, 168, 174, 188–191, 201, 203, 204–205, 206, 209, 210, 211

medical procedures, 149, 196–201

occupational preferences, 136, 142, 149, 152–153, 168, 184–185, 210

partner preferences, 136, 146, 151, 159, 182–183

physical attractiveness, 135, 141, 180–183, 188–189, 198

prostitutes/prostitution, 151, 183–185, 187–188, 211

sexual preferences and practices, xi–xii, 136, 148, 152–154, 156, 178, 211

she-males, 186–188, 190

shoplifting, 184, 185

social transition prior to surgery, 201–205

socialization between types, 147, 177–178, 211–212

voice/speech patterns, 72, 197–198, 201

Transvestites, 155, 160–161, 165, 218

Tula, 180

Twilight of the Golds (play), 113

𝒰

Ulrichs, Karl, xi

University Medical Center (Utrecht), 32

University of California at Los Angeles, 24

University of South Carolina, 24

𝒱

Visual sexual stimuli, 94

W

Weinrich, James, 84

Whitam, Fred, 135–136

Wilson, E. O., 117

Witelson, Sandra, 60

Wongprasert, Thanaporn, 135

𝒵

Zucker, Ken, 28–34, 62, 170, 178–179